MW00812203

Messerschmitt · Bf 109

Willy Radinger • Walter Schick

# Messerschmitt Bf 109

## The World's Most Produced Fighter From Bf 109 A to E

Schiffer Military History
Atglen, PA

**Cover Artwork by Steve Ferguson**

Book Design by Ian Robertson.


Library of Congress Catalog Number: 99-62935.



Translation from the German by David Johnston.

This book was originally published under the title *Messerschmitt Me 109: Das meistgebaute Jagdflugzeug der Welt Engwicklung, Erprobung und Technik, Alle Varianten: von Bf (me) 109A bis 109E* by Aviatic Verlag.

Printed in China.
ISBN: 0-7643-0951-X

We are interested in hearing from authors with book ideas on military topics.

Published by Schiffer Publishing Ltd.
4880 Lower Valley Road
Atglen, PA 19310 USA
Phone: (610) 593-1777
FAX: (610) 593-2002
E-mail: Schifferbk@aol.com.
Visit our web site at: www.schifferbooks.com
Please write for a free catalog.
This book may be purchased from the publisher.
Please include $3.95 postage.
Try your bookstore first.

In Europe, Schiffer books are distributed by:
Bushwood Books
6 Marksbury Road
Kew Gardens
Surrey TW9 4JF
England
Phone: 44 (0)181 392-8585
FAX: 44 (0)181 392-9876
E-mail: Bushwd@aol.com.

Try your bookstore first.

# Contents

## Dedication

I was associated with the Bf 109 from the first stroke of the pen until the test flights to determine its terminal diving speed. And this is how it happened: after my final examination as a machine construction engineer in 1931 and my subsequent employment in a textile machine factory, in 1934 I moved to the Bayerische Flugzeugwerke at Augsburg-Haunstetten. There I was initially employed in the design bureau, "wing department," under the direction of Dipl.-Ing. Robert Prause. It was there that I first came into contact with the Bf 109, which was then being worked on under Project Number P 1034. Among other things I designed the wing cutouts for the retractable main undercarriage. Then from 1937 I was active as group leader in the preparation bureau, where I was involved exclusively with the development of the Messerschmitt variable-pitch propeller (Me P 6, Me P 7, Me P 2). From October 1937 up to and including June 1938 I earned my military pilot's license at the "Flight Practice Station Munich-Oberwiesenfeld."

Back to Messerschmitt—the BFW GmbH became the Messerschmitt AG in 1938. At first I continued working on the variable-pitch propellers, but I also conducted individual and endurance trials with these. Because of this the management moved me to the flight test department as an engineer-pilot. Since I had amassed a large number of hours testing propellers, virtually all of the flight tests associated with the development of this type were given to me. Through these many and varied missions I came into intimate contact with the Bf 109 and its variants. Approximately 80% of all my flights were in the Bf 109. As well, I made similar test flights with the Bf 108 (propeller), Bf 110, Me 210, Me 309, Me 410 and toward the end of the war also in the twin-engined Me 262 jet.

I remember one test flight, which will be the subject of its own chapter, especially well: determining the terminal dive velocity of the Bf 109 (G-series). Carrying out this task was the high point of my flying activities in general, and with the Bf 109 in particular.

I wish this Bf 109 book many readers, especially younger ones who have grown up in the "jet age," so that they might gain some idea of what "propeller flying" was like back then.

Augsburg, 12 June 1996
Lukas Schmid

# Foreword

There is a long story behind this book. After having compiled the successful *Me 262* and *Messerschmitt Secret Projects* with Walter Schick, we planned as our next work a comprehensive history of the development, technology and testing of the Bf 109. Because of the plethora of material that is available on this most-built fighter aircraft in history and its many variants, we decided—in consultation with Aviatik Verlag—to produce this in two volumes. The first volume was supposed to deal with the story of the creation of the Bf 109 and subsequent versions up to the Bf 109 E, while a second volume would deal with the Bf 109 F through K and those variants built in Spain and Czechoslovakia after the Second World War.

Our work began in 1994, but then Walter Schick became seriously ill and died in mid-1995. In the months and weeks before his death he worked intensively on a concept for the organization and preparation of the first volume, for in all of our years together German aviation history had become his passion. Walter Schick was dedicated to accurate research and believed that the facts should be allowed to speak, and that many errors that have crept into aviation books over the years should be corrected. As a result he earned an excellent reputation among aviation historians and readers, and at this point I would like to thank Walter Schick for our years of intensive and successful cooperation. He will not be forgotten. Several experts have contributed to the completion of the manuscript, and I would like to thank them here. Several Messerschmitt employees have also contributed information to supplement the manuscript, and Dieter Herwig provided photos and helped with the text. Günther Sengfelder made an important contribution to the illustration of this book with his excellent drawings of various versions of the Bf 109. Theodor Mohr proofread the manuscript, provided drawings and gave valuable hints. Not last I would like to thank Lukas Schmid, who carried out many important test flights in the Bf 109 as a Messerschmitt company pilot, for his dedication.

But my very special thanks go to Hanfried Schliephake. Not only did he assist in the difficult task of selecting the many photographs, but he also reviewed the manuscript and helped clarify many still unanswered questions. His assistance was especially welcome in the rather hectic weeks before the book was printed.

With the support of many this book has become a factually precise and historically interesting chronicle of the first variants of the Bf 109. True, there are already many publications about this legendary German aircraft, but I believe that this work corrects many mistakes of the past and also brings to light some interesting facts for the first time.

Neusass, summer 1997
Willy Radinger

# Introduction

## Fighter Aircraft Development at the Start of the 1930s

Toward the end of the First World War a new type of aircraft began entering service with several German *Jagdstaffeln*, albeit in small numbers. Just as the world's first jet fighter, the Me 262, would do in the Second World War, it pointed the way to the future of fighter aircraft design. In 1918, long before the ultimate step was made from braced biplanes of mixed construction to all-metal, low-wing fighters, an aircraft was flying which clearly differed from every fighter aircraft yet to see service. Outwardly it was already very close to the next generation of fighters, lacking only a retractable undercarriage and enclosed cockpit.

The man who translated his concept of the modern aircraft, in this case for military purposes, into fact was none other than the brilliant Professor Hugo Junkers. The aircraft in question was the Junkers J 9, derived from the world's first all-metal aircraft, the Junkers J 1 (mid-wing monoplane of sheet steel construction). The most striking difference between the J 9 and other fighter aircraft, including those of other nations, was the fact that it was a low-wing cantilever monoplane of all-metal (corrugated Dural) construction. The following specifications may be used for comparison with later German fighter aircraft:

**In 1917-18 the Junkers J 9, an all-metal, low-wing cantilever monoplane, was a futuristic design, however, it was not pursued.**

| Type | Junkers J 9 |
|---|---|
| Year built | 1918-18 |
| Power plant | 1 x BMW IIIa |
| Takeoff power | 260 H.P. at 1,580 rpm |
| Wingspan | 9.00 m |
| Length | 7.25 m |
| Height | 2.60 m |
| Wing area | 14.12 $m^2$ |
| Empty weight | 654 kg |
| Takeoff weight | 840 kg |
| Max. wing loading | 59.5 $kg/m^2$ |
| Maximum speed | 240 kph |
| Cruising speed | 200 kph |
| Rate of climb at ground level | 6.4 m/sec |
| Service ceiling | 7 500 m |
| Armament | 2 x LMG 08/15 (7.9 mm) |

At first, however, Professor Junkers' ideas, at least those concerning fighter aircraft, did not fall on fruitful ground. Almost two decades after the World War the braced biplane still dominated in the field of fighter aircraft, even in the resurgent German aviation industry of the Weimar Republic.

A change in the course of fighter development first became evident at the start of the 1930s. This was preceded by new and sweeping discoveries in the field of aerodynamics and attempts to create new techniques and technologies for light yet stable methods of construction.

First used in racing and record-setting aircraft, in the late 1920s and early 1930s these new techniques also found their way into military aircraft design. Certainly the first to point the way in this direction were the highly-developed racing seaplanes which competed for the Schneider Cup, aircraft such as the Italian Macchi M.C. 52 (1928) or the British Supermarine S.6 (with Rolls-Royce R engine), which on 12 September 1929 established a speed record of 575.700 kph over a three kilometer course while being flown by Olebar. The paramount design principle of these aircraft was to achieve speed combined with the necessary structural strength and optimal aerodynamic form.

Another important influence on the modern single-seater—at first for peaceful purposes only—was undoubtedly the "National Air Races" held in Los Angeles, California. These contests were strictly for land-based aircraft. The Hughes Special 1B (H-1), which on 13 September 1935 set a world speed record of 567.115 kph over a 3 kilometer course, represented a certain high point in this development work.

The design methods of Americans Richard Palmer and Howard Hughes already displayed all the features of modern, high-performance aircraft construction. The methods of increasing flight performance were no longer restricted to merely raising engine output, but now also included reducing damaging drag by eliminating non-lift-producing components, retractable undercarriages, streamlined engine cowlings, and not least a method of construction which resulted in very smooth external surfaces.

In Germany it was principally the Heinkel designers Walter and Siegfried Günter, and the technical director of the Bayerische Flugzeugwerke GmbH, Dipl.-Ing. Willy Messerschmitt, who attracted the most attention with their modern designs. And it would be these same men who later entered a fierce competition to win the world speed record for their companies.

But in Eastern Europe, too, there were practical forays into this area. One such machine was the Polish PZL 11, a braced, shoulder-wing monoplane which first flew in September 1931. Advanced design principles were already evident in this machine. In any case, the PZL P-11 reached a maximum speed of 320 kph with a 200-H.P. motor. This exceeded the performance of comparable aircraft of the day, such as the American P-26 and the French Dewoitine D.510, both of which had fixed undercarriages. The modern British low-wing fighters (Hurricane and Spitfire) were still on the drawing boards.

The elegant Heinkel He 51 was representative of the classical mixed construction biplane.

In the Soviet Union, too, designers were using advanced design methods. In May 1933 a low-wing cantilever monoplane with an all-metal, monocoque fuselage and a retractable undercarriage made its first flight. The project designation was ANT-31, and the machine carried a heavy fighter armament. The responsible designer was P.O. Sukhoi, who later became known the world over. With a 500-H.P. engine, however, the machine was underpowered. More powerful engines were installed in subsequent prototypes, making possible a speed of 414 kph and an altitude of 5,000 meters in less than six minutes. This program was abandoned after eighteen aircraft had been built, in favor of a competing design which was simpler and quicker to build in quantity, the Polikarpov I-16 (Rata).

By 1935 the cantilever, low-wing monoplane of largely all-metal construction and streamlined shape had won through as the predominant fighter design. The Curtiss Hawk and Seversky P-35 appeared in the USA. In France the Morane-Saulnier 405/406 took to the air. In the Hurricane and Spitfire, Great Britain produced two advanced, well-armed designs which later proved themselves in the Second World War. In 1935 Germany's air force was in its formative stages; that year saw the first *Jagdgeschwader* receive more modern fighter aircraft (compared to the previous Ar 65). First *Jagdgeschwader 132 "Richthofen"* was equipped with the He 51 A-1, and in March 1936 *Jagdgeschwader 134 "Horst Wessel"* received the Ar 68 F (later the improved Ar 68 E). The Ar 68 was already seen as the successor to the He 51.

Both aircraft types, braced, staggered biplanes with fixed undercarriages, displayed the typical features of fighter design in the First World War, albeit in a somewhat refined form. But they were the last German examples of this type of fighter. It should be noted that both the He 51 C (131 examples) and the Ar 68 E (3 examples) were tested for front-line suitability in the Spanish Civil War beginning in 1936. Both types were employed as training aircraft and were used in action when the situation demanded it. As well, twelve He 51 Cs were exported to Bulgaria.

Significantly, both types, powered by the same engine (BMW VI), were slower than an aircraft initially designed for civil-

| | He 51 C | Ar 68 F |
|---|---|---|
| Type | single-seater | single-seater |
| | | |
| Power plant | BMW VI 7,3 Z | BMW VI 7,3 Z |
| Takeoff power | 750 H.P. | 750 H.P. |
| Propeller diameter | 3.10 m | 3.10 m |
| Wingspan (upper) | 11 m | 11 m |
| Wingspan (lower) | 8.6 m | 8 m |
| Overall length | 9.8 m | 8.5 m |
| Height | 3.3 m | 3.3 m |
| Total wing area (aerodynamic) | 27.23 $m^2$ | 27.30 $m^2$ |
| Wing loading | 71.61 kg/$m^2$ | 74 kg/$m^2$ |
| Empty weight | 1 475 kg | 1 600 kg |
| Useful load | 440 kg | 420 kg |
| Takeoff weight | 1 915 kg | 2 020 kg |
| Maximum speed at 0 m | 330 kph | 300 kph |
| at 6 km | 275 kph | 295 kph |
| Landing speed | 95 kph | 97 kph |
| Time to climb to 6 km | 16.5 min | 16 min |
| Range at height of 6 km | 730 km (optimal) | 500 km (95% of fuel expended) |
| Main fuel tank capacity | 210 l | 200 l |
| Drop tank | 170 l | |
| Armament (for both types) | 2 x MG 17 (7.9 mm) with 500 rounds per gun | |
| | Synchronized to fire through the propeller disc | |
| Gunsight | Revi 3b (reflector sight) | |
| External stores | | 5 x SC 10 10-kg bombs |
| Configuration | Single-braced biplane with staggered wings | |
| Co on ratio | 7.3 | |
| | Z = Zenith carburetor (higher peak performance) | |
| 750 H.P. at 0 m | with liquid cooling | |
| Source: | L.Dv.311 | L.Dv.390 |
| | Data Sheet D.-Bl.655 | |

ian use. It was the He 70 *Blitz*, an aircraft that was approximately 50 kph faster than the two fighters while carrying six passengers. This performance was due to its outstanding aerodynamic shape and retractable undercarriage—one of the first to be used by a German aircraft.

In spite of all the political and economic adversity, the German aviation industry nevertheless succeeded in keeping pace with and even surpassing developments abroad, producing the Bf 109 and He 112 (see the Zurich Air Meeting 1937).

And not only that: the Bf 109 design remained in service, albeit in improved form and with ever more powerful engines, until the end of the war, by which time it had become the most produced fighter ever, with approximately 32,000 examples built.

## German Air Armaments after the First World War

The terms of the Versailles Treaty of 28 June 1919 resulted in the disbandment and destruction of the German *Fliegertruppe*. At the same time it made the existence of military aviation impossible. The terms allowed only a small army (*Reichsheer*) with an effective strength of 100,000 men. Heavy artillery, aircraft and chemical weapons were forbidden. In fact, the "*Re-*

*ichswehr*" was intended solely to maintain order inside Germany and act as a border police force.

On 1 March 1920 *General* Hans von Seeckt, head of the *Truppenamt* (TA) in the *Reichswehrministerium* (RWM), set up a special office in the *Truppenamt* under the direction of *Hauptmann* Helmuth Wilberg. Titled the "*Luftschutz-Referat*" (TAL, or air defense section), it was to handle all questions dealing with military aviation. At first it concerned itself in purely theoretical terms with the organization and training requirements of military aviation. Thus, Hptm. Wilberg, who on 20 November 1941 was killed in an air crash as *General der Flieger*, laid the foundation for the later *Luftwaffe*. At the same time the "*Fliegertechnik*" (development) section was established in the *Heereswaffenamt* (army ordnance office) under *Hauptmann* Kurt Student. It concerned itself with the technical aspects of military aviation and also pursued "development and testing," at first also only theoretically.

On 1 January 1923 air sovereignty over German territory was returned to the Reich government. That same year the RWM (Hptm. Kurt Student) extended the first feelers to the German aviation industry. Hptm. Student (WaPrüf 6 F) received data from the firms involved; the following designs were chosen and contracts were issued to Arado, Albatros, BFW and Heinkel:

## The *Reichsluftfahrtministerium*

As part of the process of establishing the *Reichsluftfahrtministerium*, effective 15 May 1933 the RWM's *Luftschutzamt* was transferred to the new RLM. Together with the former *Reichskommisariat für Luftfahrt* it formed the new ministry under the leadership of Reich Aviation Minister Hermann Göring and State Secretary Erhard Milch, the former director of *Deutsche Luft Hansa*. Thus was laid the foundation for the building up of the German air force. Finally, as a result of a decree by the new Reich Chancellor, on 1 March 1933 the *Luftwaffe* was created as the third branch of the armed services alongside the army and navy. Its supreme command authority was the RLM.

For the purposes of this history the most important of the RLM's departments was the *Technische Amt (LC)*, then under the direction of *Oberst* Wilhelm Wimmer (he was replaced by then *Oberst* Ernst Udet on 9 June 1936). In turn, the most important branch of the *Technische Amt* was *Abteilung LC II* (development). On 1 April 1934 it was headed by *Hauptmann* Dr.Ing. Wolfram von Richthofen, who in 1938 became the last commander of the *"Legion Condor."*

| | Cover name |
|---|---|
| **Arado SD 1** | |
| Home defense fighter[1] | "Heitag" |
| **Albatros L 76/77** | "Erkudista" |
| Reconnaissance aircraft | |
| **BFW Bf 22** | |
| Night fighter and reconnaissance aircraft[2] | "Najaku" |
| **Heinkel He 40** | |
| Reconnaissance aircraft (multi-purpose) | "Erkunigros" |

[1] The firms were given only general information on performance and equipment, and no differing tactical requirements. The building of these aircraft made it possible for these firms to establish a basis for military aircraft construction.

[2] The BFW type designation was M 22. Described in BFW documents as a "biplane for postal purposes" for security reasons, the machine was damaged on 6 May 1930 (cost RM 5 545.10) and totally destroyed on 14 October 1930 in the hands of Eberhard Mohnike. Total cost of the aircraft was 356,205.12 Reichsmark. The design was not one of Willy Messerschmitt's, but it was during this secret armament phase that the name of Bayerische Flugzeugwerke AG (BFW) was associated with the design of military aircraft for the first time.

From the RLM came all calls for tenders for individual designs, as well as their technical-tactical requirements, development and procurement contracts, approval to build under license, allocations of mate-

**Tactical Requirements for Fighter Aircraft (Land)**
Per L.A. 1432/33 g.Kdos. of 6/7/1933

| | |
|---|---|
| 1. Tactical purpose: | Single-seat fighter, air combat by day |
| 2. Number of engines: | 1 |
| 3. Crew: | 1 pilot |
| 4. Armament: | 2 fixed machine-guns with 1,000 rounds or 1 fixed machine-cannon (20-mm, 100 rounds) |
| 5. Radio equipment: | Specialized radio equipment for ground-to-air and air-to-air communications. |
| 6. Safety and rescue equipment: | Safety harness, parachute, oxygen system, heating |
| 7. Speed: | Maximum: 400 kph at 6 000 meters |
| 8. Range/endurance: | 1 1/2 hours full throttle flight at 6,000 meters |
| 9. Time to climb: | 17 min. to 6,000 meters |
| 10. Operating altitude range: | to 9,000 meters |
| Service ceiling: | 10,000 meters |
| 11. Airfield size: | Landing field 400 x 400 meters |
| (standard German) | approach path from height of 20 m: 400 meters |
| 12. Defensive protection: | Protected against leakage fire |
| 13. Parking (in the field): | Open air |
| 14. Other: | Flight in clouds and fog possible, the aircraft's dimensions much be such as to meet rail transport requirements (normal profile). |
| 15. General remarks: | |
| to section 3 | Best visibility for air combat |
| to section 4 | Armament installation is to take into consideration best possible field of fire and ease of operation. |
| to sections 7-11 | Performance and handling qualities: the aircraft must be capable of being mastered by every average pilot. The required flight envelope is to include:<br>The entire flight endurance at 6,000 m working altitude including takeoff and gliding approach.<br>Aircraft must be able to maintain maximum speed for several periods of up to 20 min. at 6,000 m working altitude.<br>Aircraft must be flyable to practical diving speed.<br>Assuming that the aircraft can be placed in a spin, recovery from same must be easy.<br>Aircraft must be able to fly turns in working altitude range with no loss of altitude.<br><br>Performance to be evaluated in following sequence:<br>1 Horizontal speed<br>2. Rate of climb<br>3. Maneuverability |

Staffel takeoffs and landings (up to 9 aircraft) are a possibility.
The approach path adjoins both ends of the landing field.

rial and so on to the affected companies involved in the air armaments industry.

Finally, financing and payment of issued contracts was done through *Amtsgruppe LF*. At the time of its so-called "unveiling" in March 1935, in terms of fighter aircraft the *Luftwaffe* had at its disposal the Ar 65, He 51 and Ar 68.

The RLM had been aware for some time that existing developments in fast commercial aircraft and indications of fighter development abroad had rendered these designs obsolete. So in December 1933 the RLM presented its "tactical requirements" for single-seat fighter aircraft (Secret Command Matter No. L.A. 1432/33, an expanded version of T.A. No. 918/31 Secret Command Matter of 16 November 1931) to the ministry's Aircraft Development Group LC II. At that time

*Fl.Stabs.Ing.* Diederich Christensen was in charge of the responsible section (C II/1 b "Fighter and Reconnaissance Aircraft"). At first these tactical requirements formed the basis for further technical development and served as the basis for the so-called "technical requirements."

In February 1934 the RLM issued the development contracts to Arado (Ar 80), BFW (Bf 109) and Heinkel (He 112). Focke-Wulf did not receive the documents until seven months later in September 1934 (Fw 159), because it had to be assumed that the power plants anticipated for all of the prototypes, either the Jumo 210 or BMW 116, would be available by then.

In the "Technical Requirements" the terms of tender for the same aircraft type differed little based on the manufacturing company.

# The Road to the Bf 109

## A Difficult Beginning

At the beginning of the 1930s the BFW AG ran into payment difficulties which ultimately led to insolvency. Among other reasons, this was brought about by *Deutsche Luft Hansa*'s refusal to accept delivery of ten already completed M 20 transport aircraft (because of two fatal accidents involving the type), as well as by bookkeeping errors. On 1 June 1931, therefore, the Augsburg Lower District Court initiated bankruptcy proceedings.

Design work was able to continue, albeit to a modest degree, as a result of the reactivation of the "Messerschmitt Flugzeugbau GmbH" under design bureau chief Richard Bauer. In autumn 1931 the Messerschmitt GmbH also received a development contract from the *Reichsverkehrsministerium* (RVM) for an aircraft which was to be designed and built based on the tender criteria for the *Europa-Rundflug* of 1932.

Six machines were built and were assigned the type designation M 29. Among the design bureau's other projects were the M 31, a two-seat sporting aircraft (only one example built), the P 1020 project study (M 32) for a training aircraft, and a so-called "people's aircraft" to be delivered in kit form (M 33, mock-up only). Other work included a high-speed mail plane under Project Number P 1012, which was shown in model form at the DELA (German Air Sports Exhibition 1932, Berlin), as well as the model of a long-range transport aircraft (M 34).

On 2 April 1932, after preliminary work by Rakan Kokothaki, appointed business director in 1932, and lawyer M. Merkel, an enforced settlement was reached before the Augsburg court which made it possible for the company (now BFW AG again) to resume production activity. On one hand this was a new beginning, but on the other the license production contracts meanwhile assigned by the RLM were not in keeping with the ideas of 35-year-old Dipl.-Ing. Willy Messerschmitt. Instead of developing and building modern aircraft based on his own ideas, he was initially obliged to produce other companies' designs under license for the expansion of the still "secret" *Luftwaffe*. This manufacturing under license did, however, provide the necessary funding which contributed to the financial stability of the BFW AG at that time.

For the sake of completeness there follows a list of types built under license by BFW; the information comes from LC No. 13271/135 Secret Command Matter J of 27 November 1935 "Planned Deliveries from 1 October 1935 to 31 March 1937":

| | | |
|---|---|---|
| Do 11 | 30 examples | 01/07/1934 |
| He 45 | 70 examples | 01/07/1934 |
| He 50 | 35 examples | 27/11/1935 |
| Ar 66 | 90 examples | 27/11/1935 |
| Go 145 | 115 examples | 01/07/1935 |
| Ju 87[1] | 50 examples | 01/01/1935 |
| Bf 108 | 7 examples | pre-production series 27/11/1935 |
| | 45 examples | production 27/11/1935 |
| Bf 109 | 10 examples | pre-production series 27/11/1935 |
| Bf 110 | 8 examples | pre-production series 27/11/1935 |

[1] The Ju 87 does not appear in subsequent procurement programs, and it was not built under license by BFW.

With its monocoque construction, retractable undercarriage and leading edge slats, the Bf 108 was a predecessor of the Bf 109.

**All-Metal Stressed-Skin Construction: From the Bf 108 to the Bf 109**

In September 1933 the BFW AG received a development contract from the RLM for a four-seat touring aircraft, which was to participate in the *Europa-Rundflug* at Moscow-Mokotov (25 August to 16 September 1934). Six prototypes were built, of which four took part in the flight. Initially developed under the "M number" M 37, the aircraft subsequently received the type designation Bf 108. The famous aviatrix Elly Beinhorn dubbed the V6, D-IJES, the "*Taifun*" (Typhoon), a name by which the aircraft became world famous and which was retained for the subsequent series. The following is a summary of the design features and innovations of the Bf 108 (B, C and D series).

**Wings**: Stressed-skin, single spar wing with automatic leading edge slats and landing flaps. Concerning use of the slats, an agreement was reached with the English firm of Handley Page which allowed it to use BFW's patented single-spar wing, while in return Handley Page gave permission for BFW to make use of its patented slats. The wings could be folded alongside the fuselage (one of the conditions for participation in the race).

**Fuselage**: Consisted of two shells (right and left), whereby bulkheads were eliminated as separate components in the conventional sense. Instead the ends of the metal fuselage sections were flanged to act as bulkheads. Further stiffening was provided by stringers riveted to the assembled shell components.

**Undercarriage**: Fully retractable, manually operated by a worm drive and gear segment. Cantilever, hydraulically-damped oleos (a non-retractable version was first used on the M 29), hydraulic brakes (in contrast to the compressed-air brakes used by the prototypes).

The Bf 108's maiden flight took place at Augsburg-Haunstetten on 13 June 1934 with Dipl.-Ing. Carl Francke at the controls.

With the Bf 108, Messerschmitt, who since 1930 had held a teaching position at the technical institute in Munich, had brought about a change in his work from the mixed construction style with fixed undercarriage to monocoque construction with retractable undercarriage combined with the latest aerodynamic knowledge. He thus set the stage for the next RLM assignment, a "single-seat pursuit fighter" which was to become the Bf 109.

### The "Fly-Off" and the Bf 109's Competing Designs

Before proceeding further with the Bf 109 story, a brief description of the designs which were presented to the RLM in the course of the pursuit fighter's development is in order.

Furthermore, several inconsistencies concerning the "fly-off" between the competing designs, as it is referred to in all of the postwar literature, should be corrected. Quite apart from the fact that there is no agreement as to the timing (it generally ranges from 1935 through October to spring 1936), one author has all four aircraft types subsequently beginning these flights in 1935. The location of the "fly-off" is variously given as the *E-Stelle Rechlin*, then Lübeck, and also the *E-Stelle Travemünde* (which is correct).

The demonstration flights took place at the *E-Stelle Travemünde*, which was actually responsible for seaplane testing, but which unlike Rechlin was not being used to full capacity at that time. At least this was the case in October 1935. At that time only the Bf 109 V1 (under repair) was present there. According to the development program, the Ar 80 V1 was not supposed to be delivered until December 1935, after the priority level assigned to the work was lowered at the end of 1935. The He 112 V1 was not ferried to Travemünde until December 1935, and the Fw 159 V1 was just then in the final assembly stage (first flight 30 November 1935).

The only possible authentic dates for this fly-off are February and March 1936. The Bf 109 V2, which took part in the competition and which was flown by several *E-Stelle* pilots, as well as by Generals Ernst Udet and Robert Ritter von Greim, was already slightly favored as a result of the outstanding demonstrations (aerobatics and spins) given by Dr. Hermann Wurster of BFW. In fact, only the He 112 V2 (D-IHGE) could be considered a serious competitor. However, the ultimate decision was made in favor of the Bf 109, although the He 112 was very advanced aerodynamically and also had a retractable undercarriage and was of robust construction. One serious disadvantage of the Heinkel design was its elliptical wing, which did not lend itself to mass production, being much more expensive to manufacture than the trapezoid-shaped wing of the Bf 109. Finally, the method of construction employed by the Heinkel design would have been somewhat more expensive, although in the event of war this would not have a factor. The following are the log book entries of Dr. Wurster at Travemünde in 1936:

| | | |
|---|---|---|
| **Bf 109 V2** | 21/02 | Factory flight |
| | 26/02 | Demonstration flight, aerobatics |
| | 27/02 | Dive, foremost cg position |
| | 27/02 | Spins |
| | 02/03 | Dive, rearmost cg position[1] |
| | 02/03 | Spins, rearmost cg position[2] |
| **He 112 V2** | 12/03 | Comparison flights |
| **Bf 109 V2** | 19/03 | Factory flight |
| | 19/03 | Factory flight |
| | 20/03 | Demonstration flight |
| | 21/03 | Demonstration flight |

[1] Need footnote here
[2] 21 revolutions to the right, 17 to the left

**The Arado Ar 80 V2 (first flight late 1935) with its characteristic inverted gull wing and fixed undercarriage.**

## Brief Description of the Competing Aircraft

### Arado Ar 80 V1

As mentioned previously, in February 1934 the Arado Flugzeugwerke GmbH of Warnemünde also received a development contract for the "single-seat pursuit fighter." Under project number E 200, design work began on 23 April 1934. This was the first all-metal aircraft developed by Arado. Inspection of the mock-up took place on 9 June 1934. After some minor changes approval was given for construction, which commenced on 15 June.

The aircraft, a low-wing monoplane, was designed in such a way that the wing center-section, of welded steel tube with removable metal panels, passed through the fuselage. However, this tubular construction made it impossible to accommodate a retractable undercarriage at that location. Consequently, the landing gear struts were attached to the forward spar at the point of the bend in the wing (in similar fashion to those of the Ju 87). The wheels retracted aft into the wing, turning through ninety degrees in the process, but the retraction mechanism could never be made to function flawlessly. As a result, the designers turned to a fixed, faired single-leg undercarriage; the location of the attachment points remained the same. The aircraft featured the typical Arado tail section arrangement (fin and rudder in front of the horizontal stabilizer, which made it relatively spin-proof, a concept developed by Dipl.-Ing. Walter Blume). The cockpit was not fully enclosed. Like the other competing designs, the Ar 80 was powered by an English Rolls Royce Kestrel II S, since no suitable German 20-liter motor was available. The aircraft made its first flight at Warnemünde on 26 July 1935, two months after the Bf 109.

### Heinkel He 112 V1

In February 1934 the Ernst Heinkel Flugzeugwerke GmbH of Rostock, too, received a pursuit fighter development contract (LC No. 6181/34 g.H 2a), which on 19 October 1934 was increased to delivery of three prototypes.

The aircraft was designed as an all-metal, low-wing cantilever monoplane with an outwards-retracting main undercarriage. The designers strove to give the machine the most favorable aerodynamic form and employed every possible design refinement, resulting in a high top speed. The elliptical wing was slightly gulled. The cockpit was open and was equipped with a

**The Heinkel He 112 was the Bf 109's most serious competitor. Depicted here is the He 112 V9, which was powered by a Jumo 210 C and armed with three MG 17s. It was demonstrated in Austria in November 1937 and later saw action in Spain.**

headrest. Like the other competing designs the He 112 V1 was powered by a Rolls Royce Kestrel II S. After the construction of twelve prototypes powered by various engines (Jumo and DB) and used for various test purposes, on 21 January 1938 the RLM authorized an order for an initial batch of fifty machines for export, since the aircraft was no longer being considered for use by the *Luftwaffe*.

Two prototypes (first the V6 and later the V9, equipped with the Jumo 210 and machine-guns) were sent to VJ 88 in Spain (civil war of 1936-1939) for front-line trials. The following is an extract from "Deliveries of German Aircraft and Aero Engines to Allied and Neutral States from 1 January 1936 to 30 September 1942" (GL/F 1, Az.66e.10.(VA)/Nr.6247/42 secret dated 19 December 1942) (He 112 only):

| | | |
|---|---|---|
| 1937 | Japan | 4 examples |
| 1938 | Japan | 26 examples |
| 1939 | Romania | 29 examples |
| | Hungary | 3 examples |
| | Spain | 19 examples for Nationalist air force |
| 1940 | Romania | 1 example |

A preliminary contract for forty-two He 112s was concluded between the Austrian *Bundesschatz* (federal treasury) and the German Reich. The aircraft were to be delivered by 30 June 1938, however, the annexation of Austria by Germany precluded this. Incidentally, the unit price was 163,278 *Reichsmark* per aircraft.

**Focke-Wulf Fw 159 V1**

A correction right at the outset. Messerschmitt was not the last firm to receive a development contract for the pursuit fighter, as is claimed in many publications on this theme. Instead, it was the Focke-Wulf AG of Bremen that was the last to receive a contract, in September 1934, seven months after Arado, Heinkel and Messerschmitt. This late issuing of the contract to Focke-Wulf still presents several mysteries today:

- Why was the pursuit fighter contract from the aircraft development group LC II/1b, which was then the responsibility of Christensen and Lahmann, not issued until August 1934 instead of in December 1933 as was the case with the other companies?

**The Focke Wulf Fw 159 V1, which made its first flight in spring 1935. This was the last of the competing designs to be issued a development contract.**

• Why was this design, a braced, high-wing monoplane, with the wing center-section supported against the fuselage by two N-struts and crossed bracing wires, chosen, when it was obvious from the outset that it would have little chance of success against the modern cantilever monoplanes, the He 112 and Bf 109? Furthermore, this configuration required a very complicated mechanism for the retractable undercarriage, in which the cranked undercarriage legs had to be retracted rearwards into the narrow forward fuselage. This arrangement was to cause many problems. Another feature that set the Fw 159 apart from the other designs was its use of the Jumo 210 power plant.

Three prototypes were built. A claim in a recent publication that the late issuing of the contract was attributable to Udet's "good relationship" with Kurt Tank, Focke-Wulf's technical director, must remain speculation.

# The Design Principle of the Bf 109

Neither the original "Technical Requirements" nor a copy has survived. But several of the co-designers know from memory that the pursuit fighter had to be a low-wing monoplane with removable wings. The undercarriage was therefore attached to the fuselage, resulting in a relatively narrow track.

But the shape of the Bf 109 suggests that the RLM gave Messerschmitt a relatively free hand, allowing him to design the fighter based on his own concepts. In projecting a high-performance aircraft, in this case a fighter, the following requirements had to be met as best as possible:

- Optimum performance (speed, dive and climb).
- Optimum flight characteristics: stability about all three axes, roll, aerobatics, takeoff and landing characteristics.
- Minimum possible cost in man hours and material in large-scale production through simplification of design, highest quality.
- Low production costs.

The fundamental concept of the Bf 109 must have been especially good if, in the course of ten years, it was able to survive the numerous changes and resulting variants to become, by the end of the war, the most-produced and perhaps most successful fighter aircraft of its age. Today it is no longer possible to determine to what degree the work on the Bf 108 influenced the development of the Bf 109. Surely it was possible to use many aerodynamic and static calculations, the explanation of basic questions and various new construction methods, albeit in modified form in keeping with the different purpose of the machine.

Included by way of explanation is a brief technical description, which is generally applicable to all versions of the Bf 109. It is taken from LDv. handbooks. Variations will be discussed in chapters on the individual variants.

**Design Configuration**

Single-seat, single-engined, low-wing cantilever monoplane of all-metal construction.[1]

Retractable undercarriage.
Slats and landing flaps.
Enclosed cockpit.

**1. Fuselage** (static construction)

The fuselage was of monocoque design and consisted of two half shells. The skinning consisted of simple sheet Dural, which lowered production costs. It absorbed bending and torsional loads, and was also strengthened locally by flanged ribs and stringers, which were inserted through rib cutouts. The stringers were riveted to the skin (flush rivets to achieve a smooth outer surface). The front of the fuselage was closed off by a bulkhead with a semicircular cutout for the pilot's feet and the rear part of the engine-mounted armament. Passing through the cockpit was the spar bearer, which served as an attachment point for the right and left wing spars.

**Pilot's Seat**: Designed as a bucket seat (for seat parachute) with height adjustment.

[1] The main construction material was clad (highly corrosion resistant) Dural in the form of sheets, rods, ribs, stringers, tubes, rivets, etc.

On 28 May 1935 Messerschmitt company pilot Hans-Dietrich Knoetzsch made the first flight in the Bf 109 V1, which was powered by a Rolls Royce Kestrel engine.

**Canopy Framing**: The cockpit was covered by a three-piece canopy structure, whose framework consisted of welded square steel tubes. All panels were Plexiglas. The forward section was permanently attached to the fuselage. The center section served as an entry hatch and folded to the right. It was locked from inside. The rear section was attached to the fuselage spine. In an emergency the folding hood and rear canopy could be jettisoned.

**Engine Mount**: This consisted of two bearers of welded steel tube which were attached rigidly to the fuselage at four points. The upper attachment points were attached to metal fittings, the lower ones were braced against the landing gear trusses.

**2. Wing** (static construction)

The cantilever, low-mounted wing was of double trapezoid shape in outline with slightly rounded tips (including the E variant). It was built in two sections, with each half separately removable, and was attached to the fuselage at three points.

The all-metal wing was designed with a single spar (Messerschmitt patent). The spar was designed as a double-T bearer (solid girder) and stood vertical to the fuselage's longitudinal axis. The ribs were divided into nose and end ribs, either as solid or lattice ribs. The wing was metal covered, and the wingtips consisted of Elektron sheet (a type of light alloy with a very high percentage of aluminum).

**Landing Flaps**: These were located between the ailerons and the fuselage. The framework consisted of Dural with solid ribs arranged on the spar. The remaining ribs were divided into nose and end ribs. The rib ends were attached to a Hydronalium cover strip. The flaps were entirely fabric covered and were set by means of a handwheel on the left wall of the cockpit. The maximum deflection for landing was

40°, which resulted in the ailerons being automatically deflected 12° downwards. Takeoff position was 12°.

**Slats**: The slats were located in the outer part of the wings along the leading edge. They were of monocoque construction, made entirely of Dural. Solid ribs were riveted to the slats to act as stiffeners. The slats normally lay against the wing. Only at large angles of attack were they opened automatically by the air forces.

### 3. Control Surfaces

The control surfaces consisted of the two ailerons on the wing, the vertical fin at the end of the fuselage with rudder, and the high-mounted, braced (also on the E model) horizontal stabilizer with elevators.

**Ailerons**: The control surface was made of Dural, whereby the spar and the nose planking resulted in a torsionally-stiff configuration. The control surface was fabric-covered. Static balance was provided by a streamlined mass balance mounted on a bearer arm on the underside of the aileron.

**Horizontal Tail**: The horizontal stabilizer, which was divided into two parts in the middle, was braced against the lower fuselage by two faired support struts.

Each stabilizer half consisted of two Dural half shells which were reinforced by riveted sheet metal ribs.

The elevator was also designed in two parts, attached by a shaft in the center. The framework was made of sheet Dural, with both control surface ends fitted with a horn balance in whose forward corner a cast piece was riveted as a balance weight. The control surface was fabric covered.

**Vertical Tail**: The vertical fin was also constructed of two metal halves, which were stiffened by riveted ribs. The profile was asymmetrical. The fin was attached to the fuselage at two points. The rudder framework was made of Dural sheet with solid ribs. The leading edge and the lower rudder cutout were metal-covered. A cast piece was riveted into the forward-projecting mass and aerodynamic balance at the top of the rudder. The entire rudder was fabric-covered.

### 4. Undercarriage

The aircraft was equipped with two cantilever oleos (air shock absorbers with oil damping) which retracted into the wing, and a retractable tailwheel in the fuselage. Retraction was hydraulic. Taxi angle was 15°.

The so-called landing gear truss (cast steel) was designed to attach the main undercarriage to the fuselage. It combined three functions:

- Attachment of the undercarriage leg.
- Attachment of the lower strut of the engine bearer.
- Attachment of the forward wing attachment fitting (auxiliary spar)

This landing gear truss directed the landing impact loads to a statically favorable point in the fuselage. This attachment element was initially cast, but was later forged (steel). The attachment of the undercarriage to the fuselage did not permit a very wide track (2 meters), which was to lead to fierce criticism as a result of a large number of accidents and crashes during the war, mainly in training activity.

It must be taken into consideration, however, that the flight cadets, most of whom converted from the Ar 96 to the Bf 109, possessed too little experience to master this machine from the very beginning. As well, there was the machine's tendency to swing to the left. The situation improved somewhat when the two-seat Bf 109 G-12, which could carry an instructor and a student, became available (late 1943-early 1944). This undercarriage arrangement also had certain indisputable advantages:

- It was possible to change a wing without the use of specialized equipment (cranes), which had a favorable and time-saving effect, both during manufacture and in service with the front-line units.
- The aircraft was easier to stow and transport. It could be towed on roads or airfields behind a tow vehicle to which the tailwheel was secured.

**Mainwheels**: 650 x 150 mm high-performance tires were used with well rims. Braking was provided by oil hydraulic brakes, and each wheel could be braked separately. Braking was initiated by toe pressure on the rudder pedals. Undercarriage fairings were attached to the oleos. Retraction and extension of the undercarriage was hydraulic.

**Tailwheel**: The tailwheel was also retracted hydraulically, and the tire size was 260 x 85 mm. Several variants were equipped with a non-retractable tailwheel. The tailwheel could be locked and unlocked for taxiing.

**Fuel Tank**: The fuel system consisted of the fuel tank, fittings and lines, as well as the injection system. The tank was located beneath and behind the pilot's seat. Capacity varied from variant to variant: Bf 109 A and B: 235 liters, C: 337 liters. Fuel capacity could be increased through the use of drop tanks of various sizes (beginning with the E model).

**Armament**: The various armaments will be discussed in the following chapters.

# The Start of a Long Development Series: The Bf 109 V1

It remains to be mentioned that a so-called static test airframe (designated V0) was also built in pieces and was assigned the *Werknummer* 757. It was used solely for static structural strength tests, and no further information is available.

### The Bf 109 V1, WerkNr. 758, D-IABI

After receiving the development contract (15 February 1934), on 8 March the director of the project bureau, Dipl.-Ing. Robert Lusser, discussed the first details of the company's plan with Fl. Stabs-Ing. Christensen in the RLM. Barely two weeks later, on 21 March, Dipl.-Ing. Roluf W. Lucht explained in detail to Robert Christensen and Robert Lusser the RLM's ideas concerning the pursuit fighter and asked BFW to submit a tender.

The first inspection of the provisional mock-up took place in Augsburg on 11 May 1934. The main topic of discussion was the weapons and power plant installation. There were two possible armament combinations: two fuselage-mounted MG 17s (firing through the propeller disc), plus one MG C/30 L (20-mm) engine-mounted cannon or one engine-mounted MG 17. The MG FF cannon did not become available until 1937. An installation mock-up of the anticipated power plant, the BMW 116, arrived at the factory on 1 July 1934. As a result of these discussions, Messerschmitt built a completely new, more detailed mock-up. At the same time the design bureau headed by Richard Bauer began turning the project drawings into design and construction drawings that could be used by the factory.

Inspection of the design mock-up took place on 16 and 17 January 1935, and the results were entered into a report. On or about 10 December 1934 the Testing and Prototype Department under *Meister* Moritz Asam began cutting the first metal for the V1.

On 25 April 1935 the C II department head, Dr.Ing. Wolfram Frhr. Von Richthofen, visited BFW in Augsburg during the course of an official trip. The following note appears in his memo (LC II B.Nr.7050/35 g.L. of 29 April 1935): "Will fly end of May. Shop work and workmanship make a good impression. Slat design much simplified." In the memo there is also a reference to the "special wheels" planned for subsequent prototypes of the Bf 109. The narrow, high-pressure tires were not available in time for the first flight, however, so that at least the first three prototypes had to be fitted with the significantly fatter balloon tires, which made necessary bulged fairings on the upper surfaces of the wing.

The tires were not the only source of difficulty; much more serious problems were encountered with the power plant. The Jumo 210, which was by now preferred over the BMW 116, was not yet available. This problem was solved with British help. Through the good offices of the Heinkel company the RLM was able to procure several Rolls Royce Kestrel II S engines (532-583 H.P.), which were then installed in various prototypes.

On 28 May 1935 Hans-Dietrich Knoetzsch took the new fighter into the air for the first time from the relatively small airfield at Augsburg-Haunstetten. As was standard practice, the undercarriage was not retracted. The first flight was preceded by several retraction and extension attempts with the aircraft on jacks. On account of the width of the balloon tires, the undercarriage legs and mainwheels did not completely disappear within the wing, the tires hitting the upper surface of the wheel wells. The wing skinning had to be cut out in these places and covered by a bulged fairing. Afterwards the retraction sequence proceeded smoothly. One more note: the rod welded between the undercarriage legs to brace the oleos while the aircraft was taxiing that appeared in a drawing by a well-known artist never existed and was pure invention on the part of the artist. On 5 September 1935 a member of the BAL (BFW Augsburg) took charge of the V1. After various test flights, on 15 October 1935 Knoetzsch ferried the V1 to the *E-Stelle Rechlin.* Since the machine with the R.R. Kestrel had a range of only 440 kilometers and the distance from Augsburg to Rechlin was approximately 550 kilometers, he was obliged to land for fuel at Jüterbog-Damm, which was home to part of JG 132 *"Richthofen."* It was there that the later *General der Jagdflieger* Adolf Galland saw the Bf 109 for the first time.

After arriving in Rechlin, Knoetzsch took off on the demonstration flight that had to precede the hand-over to the aircraft's new owners (*E-Stelle*). He ended the demonstration with a display of aerobatics and initiated the landing procedure. It was then that the V1's bad luck began. At this point we turn to an account by an unidentified eyewitness. He was a senior aero engineer or factory engineer who had access to the aircraft on display at any time. He wrote an eight-page "confidential report" for a likewise unidentified client (probably Heinkel), in which the aircraft's most important components were described, sketched and, in some cases, measured.

**The Bf 109 V1 on the company airfield at Haunstetten. The clean aerodynamic lines of this low-wing cantilever monoplane are evident in the two photographs above. The photo on the left illustrates the type's characteristic undercarriage, which was attached to the fuselage; this answered a requirement for removal of the wings without placing the machine on jacks.**

"The aircraft was demonstrated at Rechlin by the factory pilot and crashed during a rather hard landing. After bouncing rather too high, it came down from a height of 1 to 1 1/2 meters onto its left landing gear leg and tailwheel. The tailwheel broke off, the fuselage was bent in three places,

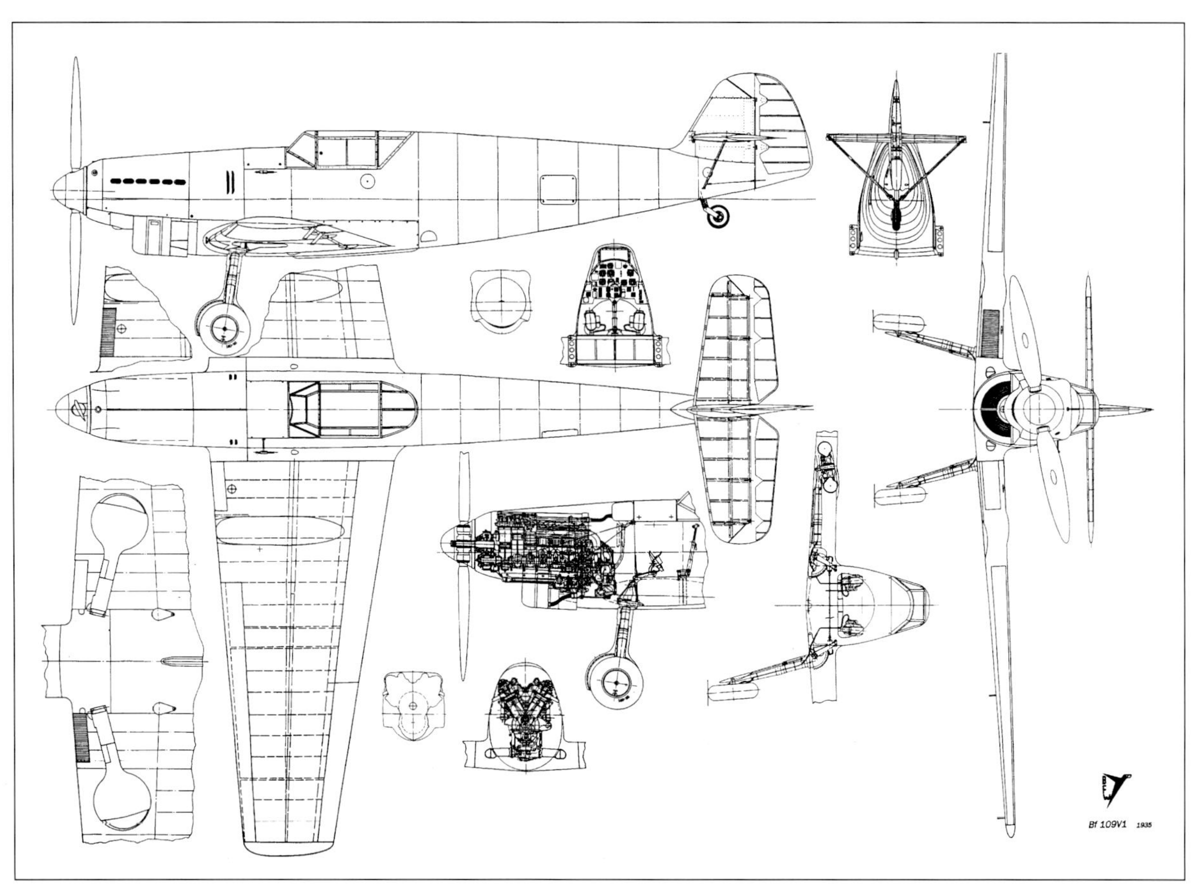
Bf 109V1 1935

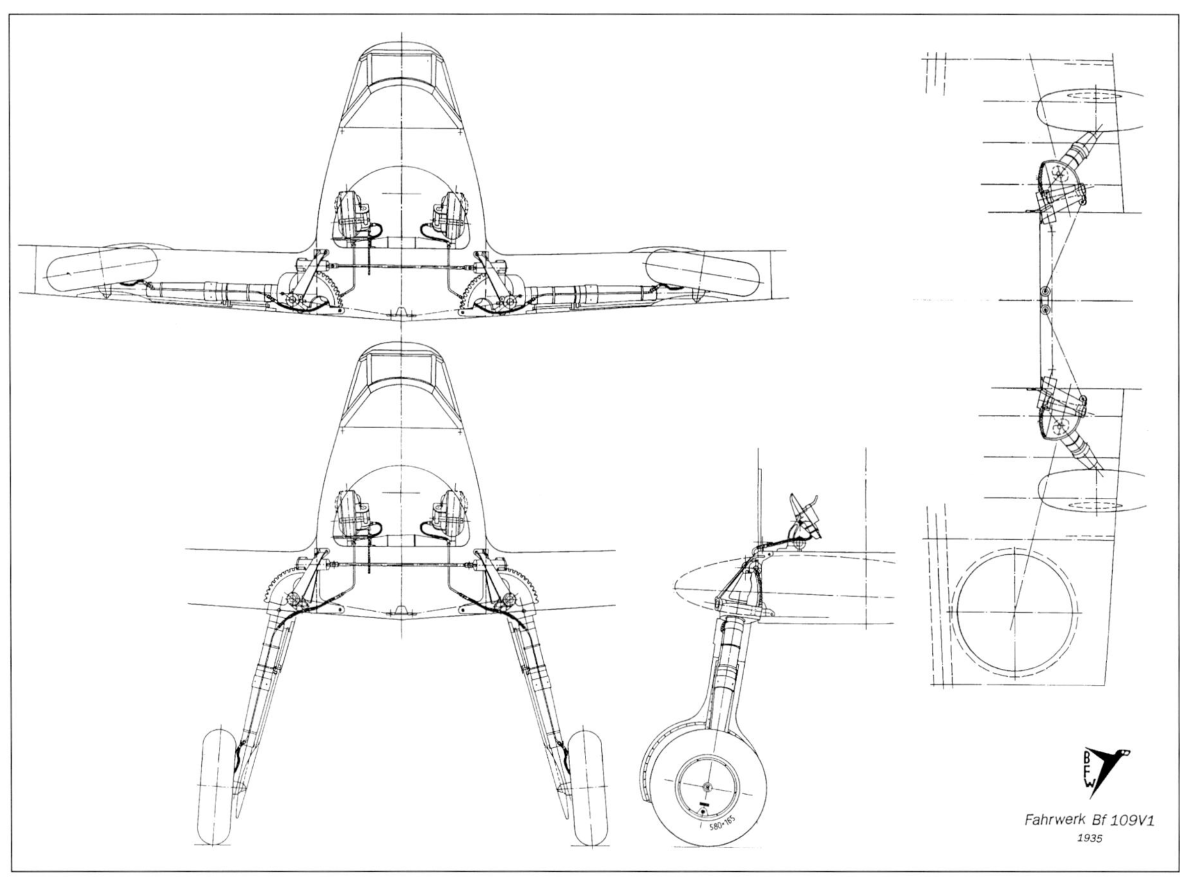
BFW
Fahrwerk Bf 109V1
1935

and the left undercarriage leg was bent inwards. The aircraft then touched down with its left wingtip, tipped over to the right and ended up on its nose. The aircraft was disassembled after this crash, and I had the opportunity to examine the individual components."

It remains to be mentioned that Messerschmitt let Knoetzsch go immediately after the crash. Since the damage proved repairable, the V1 was returned to the factory in Augsburg, where it was repaired. As a result, however, it was unavailable for testing for some time. When repairs were complete the aircraft was ferried to the *E-Stelle Travemünde*, where Dipl.-Ing. Gerhard Geicke flew it for the first time on 26 February 1936. The V1 did not participate in the so-called "fly-off," which in reality never took place in the anticipated form, for the Bf 109 V2, already powered by the Jumo 210, made its first flight on 12 December 1935. This machine was ready as "competition" for the He 112. The Bf 109 V1 was used for various test flights in Travemünde, for example 54 flights totaling 9 hours, 39 minutes in June 1936. The purposes of these flights included performance measurements, landing measurements, determination of the minimum turning radius, investigation of stall characteristics and much more.

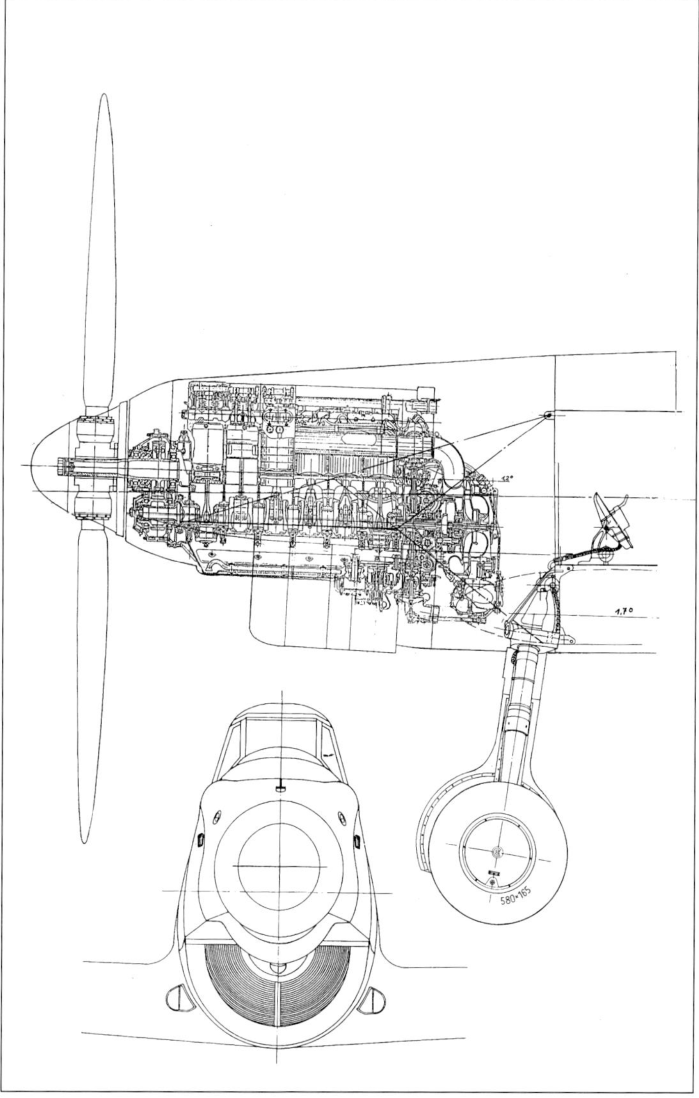

On 17 July 1936 the (new) BFW chief pilot Dr.Ing. Hermann Würster flew D-IABI back from Travemünde to Augsburg for overhaul. It was then retained by the company as a flying test-bed. It was used primarily for stall and spin tests with free slats shortened on the inner end. They were shortened in order to determine the effects of a wing-mounted weapons installation. The first of this series of test flights took place on 24 August 1936, the last recorded flight on 13 January 1937. Wendel's last log book entry appears on 11 February 1938: "Test flight, 25 min." Then all trace of the Bf 109 V1 disappears. According to the memory of a former Messerschmitt employee it was pushed behind a hangar, where it then sat unused for months. Ultimately, the first example of approximately 32,000 "progeny" was scrapped.

**General arrangement drawing of the Bf 109 V1 with details (top of facing page); drawings of the undercarriage installation (bottom of facing page); schematic section of the engine installation (above) with the Rolls Royce Kestrel. (Drawings: Günter Sengfelder)**

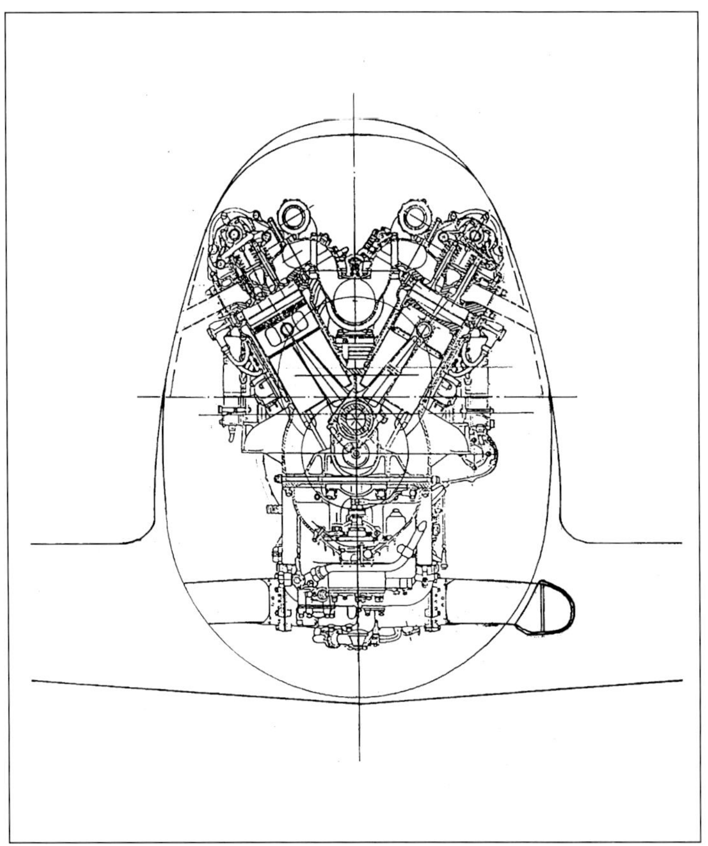

Cross-section of the engine installation, clearly showing the vee arrangement of the twelve-cylinder Rolls Royce Kestrel.

## Bf 109 V2, WerkNr. 759, D-IILU

The aircraft made its maiden flight on 12 December 1935 in the hands of Joachim von Köppen (DVL), who together with Willi Stör (DVS) was temporarily placed in charge of acceptance flying after the release of Knoetzsch. The Jumo 210 A (680 H.P. for takeoff) was installed in the machine. Arrangements were made for fuselage-mounted armament, however, the weapons were not installed.

On 21 February 1936 BFW chief pilot Dr.Ing. Würster ferried the prototype to the *E-Stelle Travemünde* to take part in the "fly-off."

D-IILU had but a short life. On 1 April 1936, during an endurance test, it made a crash landing near Ivendorf (south of Travemünde). Shortly after takeoff the windscreen flew away, compelling pilot Trillhase (*E-Stelle Travemünde*) to make a forced landing. The disassembled machine was returned to the factory and was scrapped.

**Technical Data Bf 109 V1**
(incomplete)

| | |
|---|---|
| Wingspan | 9890 mm |
| Length | 8884 mm |
| Max. fuselage width | 1160 mm |
| Span of horizontal stabilizer | 3000 mm |
| Wing area | 16 m$^2$ |

**Power plant**
Rolls Royce Kestrel II S
(S = supercharger)
12-cylinder V-engine with vertical cylinders

| | |
|---|---|
| Total displacement | 21.24 l |
| Fuel | 87 octane |

Dimensions: approximately those of the Jumo 210 (exhaust ports were at top, however)

| | |
|---|---|
| Propeller | Schwar light wood propeller |
| Diameter | 3000 mm |
| Weight of power plant | 676 kg (without propeller) |

**Performance**

| | |
|---|---|
| Maximum output | 583 H.P. at 3,000 rpm at 3 850 m |
| Continuous output | 532 H.P. at 2,300 rpm |

**Cooling System**

The semi-circular radiator is located beneath the engine. The arrangement is very well designed aerodynamically. Prior to being ferried to Rechlin for the first time the original R.R. radiator was replaced with a rather larger German radiator (11 liter capacity).
The oil cooler (skin-type oil radiator) is installed in the leading edge of the right wing near the fuselage and is equipped with thin cooling fins. The oil tank is located in the leading edge of the left wing and is accessible by removing the leading edge panel.

**Weights**
The following may be read off rear fuselage side in the side-view photograph:

| | |
|---|---|
| Equipped weight: | 1 404 kg |
| Usable load: | 397 kg |
| Takeoff weight: | 1 800 kg |
| Maximum number of persons: | 1 |
| Last check: | |
| Next check: | |
| Owner: | |
| Holder: | E-Stelle Rechlin |

Assessment of handling qualities:
Satisfactory apart from somewhat weak stability about the longitudinal axis. Directional stability to be raised by increasing dihedral:
Bf 109 V1 = 4° Bf 109 V2 = 7° 10'.
Control forces in order, control effectiveness is good.

## Bf 109 V3, WerkNr. 760, D-IOQY

The V3 was actually the prototype for the A-series, which is hardly mentioned in Bf 109 literature. Dr.Ing. Würster carried out the first flight in the aircraft on 8 April 1936. Following the crash of the V2, which had been earmarked for weapons trials, the V3 was the first armed Bf 109. Because of ongoing problems with the MG C/30 L, which was supposed to be installed as an engine-mounted cannon, the armament consisted of just the two fuselage-mounted MG 17s, which were fired mechanically at first.

On 29 June 1936 Dr. Würster flew the aircraft north of Augsburg over the Lech (see photo 32), in order to give company photographer M. Thiel an opportunity to document the only flyable Bf 109. In the photo it is obvious that the machine's propeller spinner is suitable for an engine-mounted cannon, however, no cannon was installed. Also clearly visible are the two bulges on the upper surface of the wing for the balloon tires. On account of difficulties with the MG C/30 L and the need to adhere to the delivery plan for the *Legion Condor* (VJ/88 = Experimental Fighter *Staffel* 88), the decision was made to equip the Bf 109 A with just two MG 17s (7.9-mm). Concerning the So-3 equipment, it remains to be said that the V3 and other Bf 109s were equipped with an electrically-operated vertical magazine for five 10-kg bombs (*Elvemag 5 C X*). This was installed behind the pilot's seat and the fuel tank. Since the electric fuses for the SC 10 were not yet available, it would only have been possible to drop the bombs from low altitude. A directive was issued not to install the magazine in further Bf 109s. Thus, the So-3 trials were dropped. On 1 July 1936 Dr.Ing. Würster ferried D-IOQY to Travemünde for further testing.

Several examples of tests carried out in September 1936: So-1 test (fixed arma-

**Installation of the Jumo 210 A in the Bf 109 V2 with belly radiator (above). The Jumo 210 A's welded steel tube engine bearer was resiliently attached to the fuselage attachment points. Clearly visible are the two MG 17 machine-guns mounted above the engine (left).**

ment), radio tests, performance trials, roll tests, endurance trials as part of the "especially urgent" type testing (for operations in Spain).

Still pending for October: demonstration on 4 October 1936 on the occasion of the harvest festival on the Bückeberg and in Staaken, So-3 (air-dropped weapons) and power plant trials. Endurance trials up to 300 takeoffs. In September 1936 the V3 completed 36 flights totaling 16 hours and 54 minutes. It displayed several innovations: the new FuG VII by Telefunken, a short-wave radio set for air-to-air and air-to-ground communications, was installed for the first time. The set was operated by a transmit button located on the KG 12 B (Argus) control column grip. Further: two MG 17s were installed above the engine on a common mount, lying staggered and side by side, synchronized to fire through the propeller disc. The ammunition (flexible belt) was housed in two ammunition boxes each containing 500 rounds. The aircraft was powered by a Jumo 210 C (approximately 700 H.P. for takeoff), which drove a fixed-pitch, wooden propeller (Schwarz). The oil cooler was located beneath the left wing near the fuselage.

**Following the crash of 1 April 1936, the Bf 109 V2 was initially used to test a modified canopy structure and was later scrapped (above). The Bf 109 V3 during a photo flight over the Lech River north of Augsburg on 29 June 1936; pilot was Dr. Hermann Würster.**

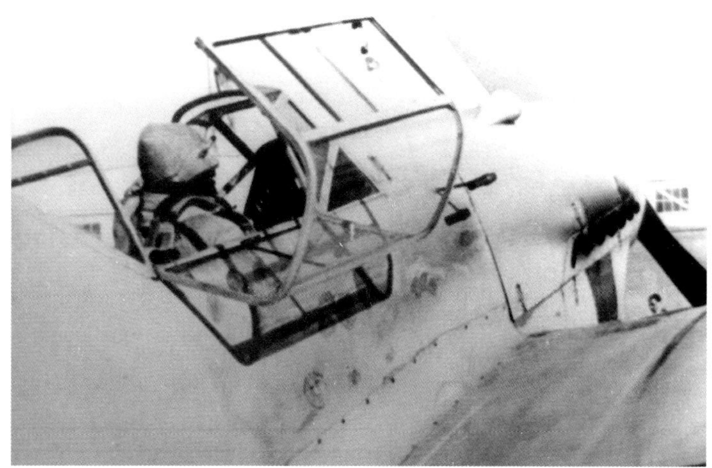

**Ernst Udet flew the Bf 109 V2 in January 1936 (above left and top of page). Arrival and unloading of the first Bf 109 in Tablada Spain for the experimental unit VJ/88 (left and above).**

**Summary of the Bf 109 A-Series Based on Messerschmitt Documents**

**Bf 109 A**

| WerkNr. | Registration | Power plant | First flight | Delivery | Fate |
|---|---|---|---|---|---|
| 760V3 | D-IOQY | Jumo 210 C | 08/04/1936 | 30/06/1936 | Travemünde, 12/36 to Spain |
| 808 | D-IIBA | Jumo 210 D | 31/12/1936 | 21/01/1937 | Rechlin, 30% damage in crash on 7/4/1937 |
| 809 | D-IUDE | Jumo 210 D | 08/01/1937 | 19/02/1937 | not known[1] |
| 810 | D-IHNY | Jumo 210 D | 08/01/1937 | 21/01/1937 | Rechlin |
| 883 | D-ITGE | Jumo 210 D | 31/12/1936 | 14/01/1937 | not known |
| 884[2] | D-IXZA | Jumo 210 D | 30/12/1936 | 19/01/1937 | Rechlin. 1/4/1938 at factory for trials |
| 994 | D-IMRY | Jumo 210 B | 06/01/1937 | 01/02/1937 | Berlin-Tempelhof |
| 995 | D-IPLA | Jumo 210 B | 09/01/1937 | 19/02/1937 | not known |
| 996 | D-IVSE | Jumo 210 B | 08/01/1937 | 19/02/1937 | not known |
| 997 | D-IZQE | Jumo 210 D | 12/01/1937 | 14/01/1937 | not known |
| 1000 | D-IMTY | Jumo 210 D | 22/01/1937 | 19/02/1937 | not known |
| 1001 | D-IPSA | Jumo 210 B | | | Total write-off on 26/02/1937 |
| 1002 | D-IQMU | Jumo 210 B | 28/01/1937 | 18/02/1937 | not known |
| 1003 | D-IVTO | Jumo 210 B | 28/01/1937 | 20/02/1937 | not known |
| 1004 | D-ILZY | Jumo 210 B | 02/02/1937 | 19/02/1937 | not known |
| 1005 | D-IJFY | Jumo 210 D | 05/02/1937 | 20/02/1937 | not known |
| 1006 | D-IBLE | Jumo 210 D | 04/02/1937 | 20/02/1937 | not known |
| 1007 | D-IHDU | Jumo 210 D | 10/02/1937 | 20/02/1937 | not known |
| 1008 | D-IYTY | Jumo 210 D | 12/02/1937 | 20/02/1937 | not known |
| 1009 | D-IOMY | Jumo 210 D | 16/02/1937 | 20/02/1937 | not known |

[1] To VJ/88 in Spain by sea. All aircraft were equipped with So-1 and So-3 (Elvemag), however the bomb magazine was not used. The two MG 17s were still mechanically operated.
[2] Was the original V10.

**Partial Data on the Bf 109 A Series**

| | |
|---|---|
| Wingspan | 9.9 m |
| Overall length | 8.7 m |
| Maximum height, tail down | 2.5 m |
| Wheel base | 2.0 m |
| Aerodynamic surface | 16.40 m$^2$ |

**Jumo 210 D Engine Performance**

| Height | Short-term output at 2,700 rpm | Raised continuous output at 2,600 rpm | Continuous output at 2,510 rpm |
|---|---|---|---|
| Km | H.P. | H.P. | H.P. |
| 0 | 680 | 610 | 545 |
| 2.7 | 640 | 575 | 510 |
| 4.5 | 500 | 450 | 400 |
| 6.0 | 406 | 365 | 325 |

As part of front-line service trials, at the end of October 1936 the V3 was packed in crates and shipped by sea to VJ/88 in Spain. The experimental unit was part of the fighter units within the *Legion Condor*, which was stationed at Tablada near Seville. The Bf 109 V4 and V6 accompanied it for service trials.

## Bf 109 A Series

The A-series was similar in almost every detail to the following B-series. The only difference was the A's lack of equipment for an engine-mounted cannon. The Jumo 210 B or D powered almost all examples of this variant. All A-series machines were built in Augsburg. A partially incomplete table titled "Delivered Bf 109 As and B-1s" dated 25 August 1937 was used to compile the above list of Bf 109 A aircraft, most of which went to the Legion Condor in Spain. According to delivery program No. 8 of 15 August 1938 (Nr. 138/38 g.Kdos.) twenty-two aircraft had been requested and delivered. According to another delivery plan only twenty aircraft were delivered.

**The Bf 109 V4, which was later sent to Spain (above). Assembling the first Bf 109 B-2s in Spain (left). Bf 109 B-1 in Spain (below).**

## Bf 109 V4, WerkNr. 878, D-IALY

This aircraft was equipped like the V3. So-1 armament installation above the engine, mechanical firing. It served as prototype for the B-series and was powered by the Jumo 210 B. The aircraft made its maiden flight on 23 September 1936 with Dr. Würster at the controls. After a few test flights, on 30 September he ferried the machine via Gotha to the *E-Stelle Travemünde*, to which it was only "temporarily assigned" since it was already scheduled to be shipped to Spain in December. No testing was begun, which meant that it was not under any circumstances flown by the *E-Stelle* in October 1936. The V4 reappears in Augsburg on 6 November 1936 when it was taken up on a factory test flight by Dr. Würster. On 30 November it was called away for a "special purpose" (Spain).

The Hawker Hurricane (above) and Supermarine Spitfire (right) were the British opponents of the Bf 109 (far right), especially during the Battle of Britain. All three aircraft shared a similar basic design, being low-wing cantilever monoplanes. From the beginning, however, the British fighters were much better armed than the Bf 109.

**Bf 109 B-Serie (Kurzbeschreibung nach LDv. 557)**

| | | |
|---|---|---|
| **Bauweise** | Siehe unter »Das Konstruktionsprinzip der Bf 109« | |
| Abmessungen | Spannweite | 9,90 m |
| | Länge über alles | 8,70 m |
| | Höhe (Sporn am Boden) | 2,45 |
| | Spurweite | 2,00 m |
| | Flächeninhalt | 16,00 $m^2$ |
| | Flügelprofil | NACA 2 R |
| | Rumpfquerschnittfläche max. | 0,9 $m^2$ |
| | Triebwerk: Siehe unter Bf 109 A: | Jumo 210 D |
| | Kühlung: Wasser | Bauchkühler unter Rumpfvorderteil |
| | Öl | Lamellenkühler unter der linken Fläche. |
| **Gewichtsübersicht** | Rumpf | 140 kg |
| | Fahrwerk | 144 kg |
| | Leitwerk | 46 kg |
| | Steuerwerk | 21 kg |
| | Tragwerk | 209 kg |
| Flugwerk | | 560 kg |
| | Triebwerk komplett | 797 kg |
| | Ständige Ausrüstung | 75 kg |
| Leergewicht | | 1432 kg |
| | Zusätzliche Ausrüstung | 148 kg |
| Rüstgewicht | | 1580 kg |
| | Kraftstoff (235 Ltr.) | 177 kg |
| | Schmierstoff (24,5 Ltr.) | 22 kg |
| | Besatzung | 100 kg |
| | Nutzlast | |
| | (1600 Schuß gegurtete Munition) | 52 kg |
| | Gepäck | 15 kg |
| | ESK 2000[1] | 9 kk |
| Fluggewicht | | 1955 kg |
| Leistungsangaben | Fluggewicht max. | 2000 kg |
| | Geschwindigkeit max. | |
| | (Waagerechtflug in Bodennähe) | 446 km/h |
| | Sturzfluggeschwindigkeit max. | 800 km/h |
| | Geschwindigkeit max. | |
| | (ausgefahrene Landeklappe) | 250 km/h |
| | Landegeschwindigkeit | 105 km/h |

Flächenbelastung 122,5 kg/$m^2$
(bezogen auf 5 min. Kurzleistung 640 PS in Volldruckhöhe 2,7 km, Fluggewicht 1900 kg)

Flugleistungen (bei Fluggewicht 1900 kg und VDM-Verstellschraube)

**Geschwindigkeiten und Steigzeiten**

| Höhe km | $V_{max}$ km/h | Steigzeit von 0 bis min | $V_{reise}$ km/h | Bemerkungen |
|---|---|---|---|---|
| 0 | 430 | | 371 | bei höchstzulässiger Dauerleistung |
| 1 | | 1,25 | | |
| 2 | | 2,58 | | |
| 2,7 | 460 | | 395 | |
| 3 | | 4,00 | | |
| 4 | | 5,58 | | |
| 5 | | 7.50 | | |
| 6 | 428 | 9, 80 | | |
| 7 | | | | |
| 8 | | | | |
| 8,75 | | | | Dienstgipfelhöhe |

Flugdauer: 1,2 Std. Vollgasflug in 6000 m.

[1] ESK 2000 = Elt.betriebene Schmalflilm-Kamera (16 mm Film) für ca. 2000 Schußbilder. Belichtungszeit: 1/150. Sie wurde für Luftkampfübungen eingesetzt und war im Bedarfsfalle auf der linken Tragflächenoberseite (1080 mm von Rumpf-Mittellinie) befestigt.

# Bf 109 B-Series

The aircraft was a "light fighter aircraft." Armament consisted of three fixed MG 17 machine-guns in the forward fuselage with electric-pneumatic remote operation. Usage group H. Stress group 5.

## Bf 109 B Weapons Installation

Two MG 17s were synchronized to fire through the propeller disc, the third was unsynchronized and fired through the hollow spinner. The two unsynchronized weapons were mounted in a staggered arrangement on a common mount allowing the right machine-gun to be fed from the left and the left machine-gun from the right. The engine-mounted machine-gun was fed from the right. The articulated ammunition belts (500 rounds per synchronized gun and 600 rounds for the engine-mounted machine-gun) were housed in three ammunition boxes; empty cartridges and belts were fed into the lower portion (belt and shell casing box). Access panels

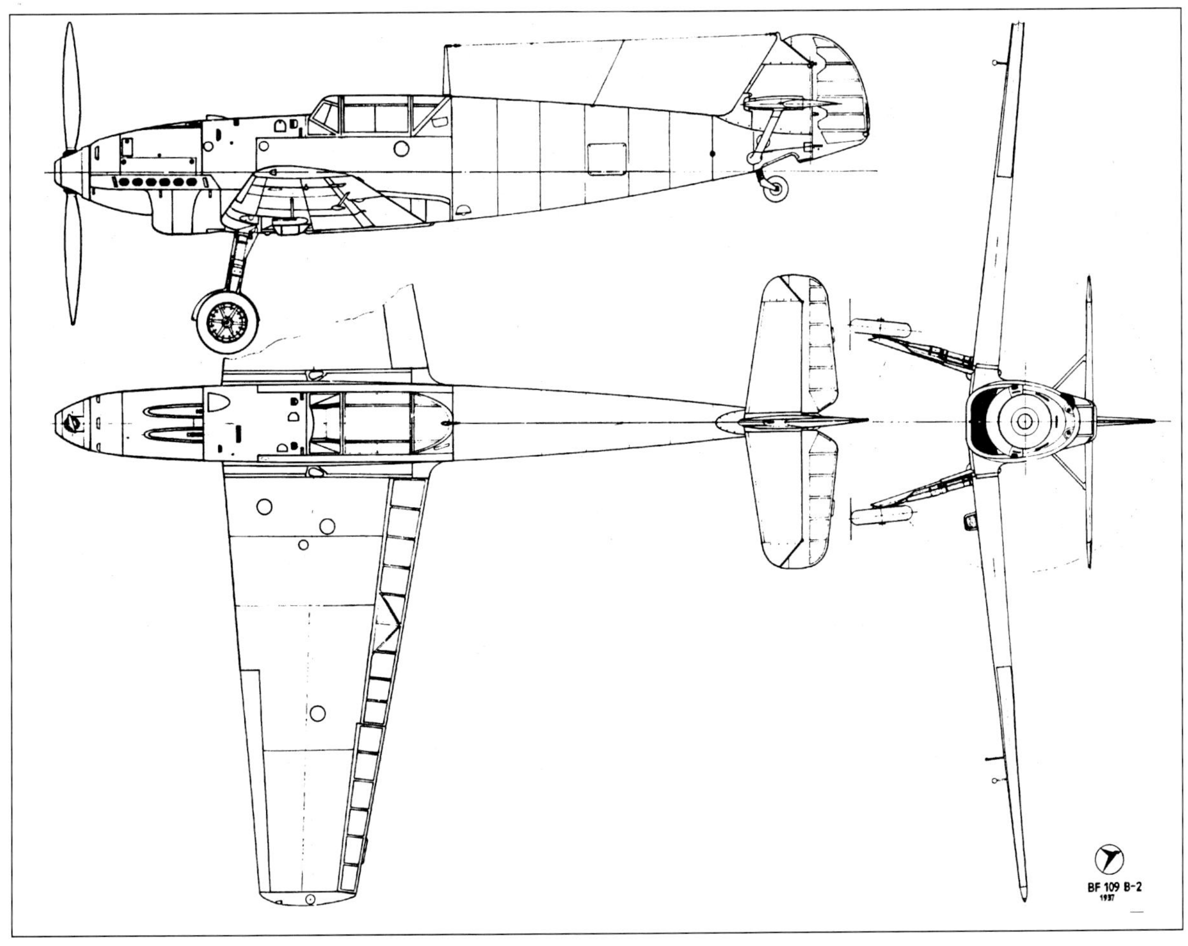


**General arrangement drawing of the Bf 109 B-2 (Günter Sengfelder).**

Bf 109 B-2 at Zurich-Dübendorf (top of page). One of the first Bf 109 B-2 production aircraft (above and right).

in the fuselage made it possible to install and remove the ammunition boxes as well as empty them.

Cocking and firing was by means of the EPAD 17. Firing was electric, while cocking was accomplished using compressed air (three bottles each containing 1 liter of compressed air at 150 atmospheres, which was reduced to 25 atmospheres operating pressure). Aiming was by means of a Revi C/12B or C/12C reflector gunsight, which was installed in front of the pilot at eye level. The ESK 2000 mentioned in the table of weights was used mainly for training purposes and was mounted on the upper surface of the left wing.

The engine-mounted MG 17 was mounted in the *Mol. 102-17* (*Mol. = Motorlafette*, or engine gun mount). A shield tube passed through the spinner and hollow propeller shaft to the rear machine-gun mount; air entering the tube also served to cool the barrel (Ldv. 288/1 Fixed Armament). The tendency of the engine-mounted machine-gun to jam resulted in the entire installation being removed from most Bf 109 Bs; this certainly reduced weight, but it also lessened the number of hits. This was clearly demonstrated during operations in Spain. With their heavy wing armament, the new British fighter aircraft (Spitfire and Hurricane) possessed a clear advantage in firepower over the Bf 109. This resulted in the design by Messerschmitt of a new "gun wing," which could accommodate either two MG 17 machine-guns or two MG FF

Several Bf 109 B-2s in the Spanish Civil War.

**A Bf 109 B-1 of 2.J/88 (code 6·15) serving in Spain during the civil war (above). A Bf 109 B-1 with the typical belly radiator of the Jumo 210 D (right).**

cannon, increasing total armament to four weapons. The Bf 109 V11 and V12 served as test-beds for the new wing.

At this time the Augsburg factory was operating at peak capacity, building the Bf 108, Bf 109 and Bf 110. Consequently, the remaining B-series aircraft were built at Messerschmitt's Regensburg factory. Furthermore, the RLM issued license contracts to the Gerhard Fieseler Werke GmbH in Kassel and the Erla Maschinenwerk GmbH in Leipzig, which by the end of May 1937 had already completed seventeen Bf 109 B-1s.

Despite claims to the contrary, company records clearly reveal that the B-1 was the only variant of the B-series. According to LC 7/1 Nr. 183.8/38 of 15 August 1938, production of the Bf 109 B-1 reached the following totals by 31 May 1938:

| | Proposed | Actual |
|---|---|---|
| BFW | 76 | 76 |
| Erla | 175 | 175 |
| Fieseler | 90 | 90 |
| | — | — |
| | 1 | 341 |

(Equal to the total number of Bf 109 B-1s delivered)

35 aircraft were sent to Spain with fuselage codes 6•7 to 6•45.

The following is an extract from a document concerning series production and retrofitting with three MG 17s:

LC II 5 u. 3
Nr. 1320/37 II 5a geh. of 1 March 1937
Subject: Bf 109 B-1 series with 3 MG 17

a) Series production of the aircraft with three electrically-operated MG 17s is proceeding according to plan in March production by the firms
BFW from WerkNr. 1021
Erla from WerkNr. 299
Fieseler from WerkNr. 3021 (target from 3016)

b) To be retrofitted with three MG 17s are the following aircraft which were completed or are under construction with two MG 17s in order to fulfil the delivery plan:
BFW from WerkNr. 1010 = 11 aircraft
Erla from WerkNr. 284 = 15 aircraft
Fieseler from WerkNr. 3001 = 20 aircraft

As of 3 July 1937 the license firms Erla and Fieseler have Bf 109 WerkNr. 998 with three MG 17s as a prototype (without wings, tail and propeller).

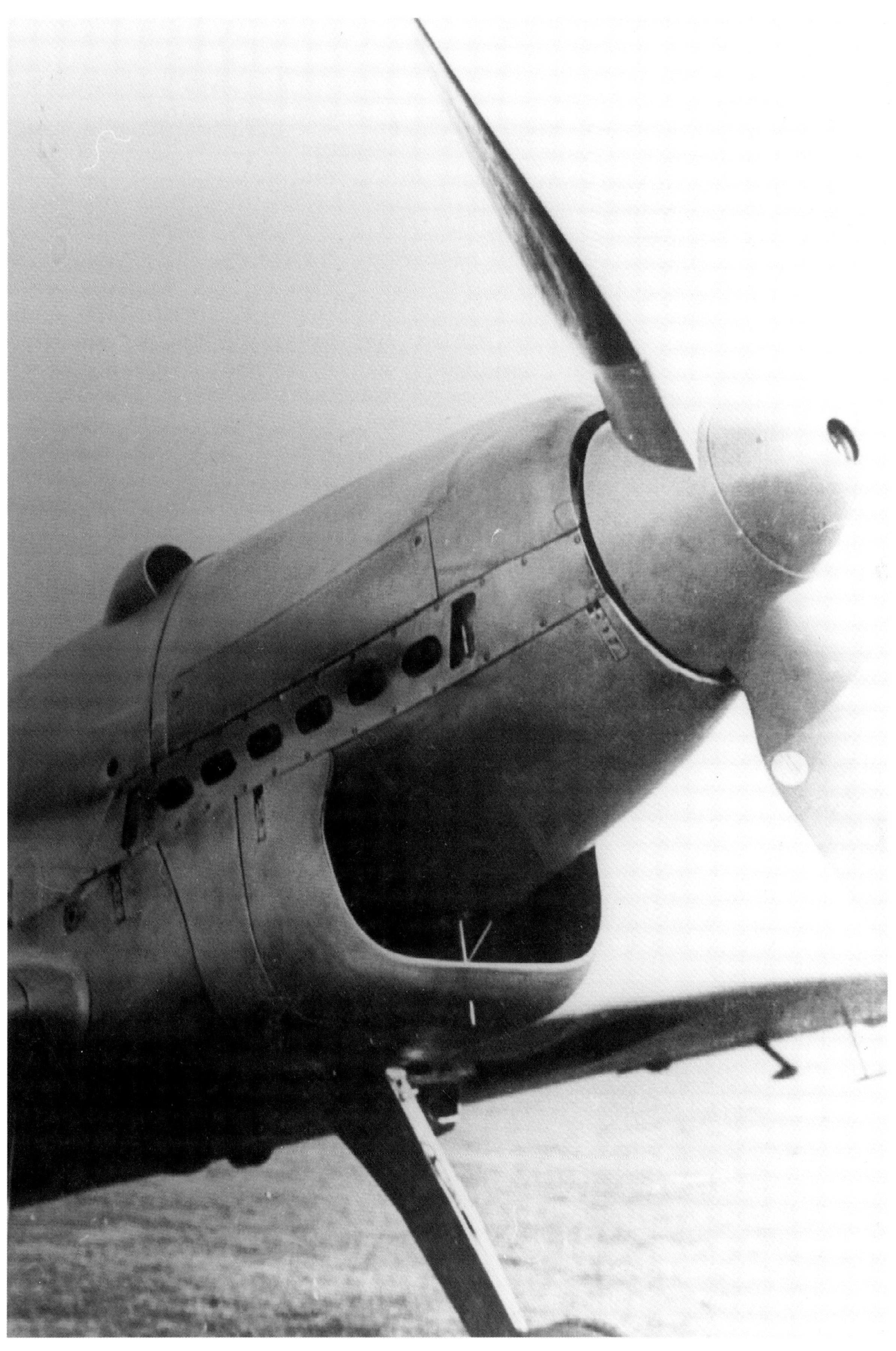

# From the Bf 109 V5 to the Bf 109 V14

### Bf 109 V5, W.Nr.879, D-IIGO

First flight by Dr. Würster on 5 November 1936. It served as test-bed for the electro-pneumatic *Abzugs- und Durchladevorrichtung 17* (EPAD 17) (firing and cocking mechanism). The aircraft was powered by a Jumo 210 B. In January 1937 the V5 was transferred to Travemünde for further testing of the EPAD 17 in test-stand and air firing. The following comment appears in the *E-Stelle*'s weekly report for the period 30 September to 7 October 1937:

| Number of rounds fired | |
|---|---|
| Test-stand: | 113,244 |
| Air firing: | 329,009 |

The V5 returned to Rechlin at the end of 1937, since beginning in 1938 the *E-Stelle Travemünde*, which had been under construction since 1935 and which among other things was responsible for testing fixed aircraft weapons, started its own testing operation.

### Bf 109 V6, WerkNr. 880

Dr. Würster made the first flight in the aircraft on 11 November 1936. The installed power plant was the Jumo 210 D (carburetor engine). Other equipment was the same as the V3. It, too, was sent to Spain for service trials, arriving there in December 1936.

### Bf 109 V7, WerkNr. 881, D-IJHA

This aircraft was earmarked for participation in the IV International Air Meeting in Zurich (23 July to 1 August 1937). It was equipped with neither the So-1 nor So-3.

**Dr. Hermann Würster in the cockpit of the Bf 109 V7 (right and bottom of facing page).**

Dr. Würster took the aircraft up on its maiden flight on 5 November 1936. It was powered by the Jumo 210 G (730 H.P. for takeoff, fuel-injected engine with two-stage supercharger, total volume 19.7 liters).

Following the air meeting the machine returned to Augsburg and under a loan agreement with the RLM served as a test-bed in a variety of experimental programs.

**Bf 109 V8, WerkNr. 882, D-IMQE**

First flight by Dr. Würster on 29 December 1936. The aircraft was equipped with a Jumo 210 D and two-blade VDM variable-pitch propeller plus two mechanically-operated MG 17 machine-guns. This machine was initially used in Augsburg for range measuring flights, after which it went to Rechlin as an engine test-bed. It is possible that at Rechlin in 1937 the V8 was fitted with the only DB 600 used by a Bf 109.

**Bf 109 V9, WerkNr. 1056, D-IPLU**

This aircraft, whose maiden flight took place on 23 July 1937 in the hands of factory pilot Fritz Wendel, was also earmarked for participation in the meet in Zurich. On 28 October 1937 it was ferried to Rechlin, where it underwent further testing.

The Bf 109 V7 after arriving at Zurich-Dübendorf for the air meet.

**Bf 109 V10, WerkNr. 884, D-IXZA**

First flight on 30 December 1936 by Dr. Würster, its 1 1/2 hour flight on 2 January 1937. The aircraft was test-flown by an RLM official on 9 January, and on 19 January went to the *E-Stelle Rechlin*. It was the original V10. Why a V10a also appears in the factory lists can no longer be explained.

**Bf 109 V10a, WerkNr. 1010, D-IAKO**

The date of its maiden flight is not known, its 1 1/2 hour flight took place on 1 November 1937. A BAL official took charge of the aircraft on 5 February, and it remained "on loan" with BFW as a test aircraft.

**Bf 109 V11, WerkNr. 1012, D-IFMO**

First flight on 1 March 1937. It was the first prototype with the new gun wing. The aircraft was powered by the Jumo 210 D.

**Armament:** Fuselage armament of two synchronized MG 17s was retained. The two 1,000-round ammunition belts were housed in two ammunition boxes located between the front fuselage bulkhead and the engine, with spent casing chutes leading to the common empty belt container beneath them. The ammunition boxes were installed and removed from the right side of the aircraft, while the spent casing and belt container was emptied through a fuselage access panel.

The two wing-mounted MG 17s were fed by continuous ammunition belts, each holding 500 rounds, running through an upper (feed) and lower belt channel over two sprung belt rollers in the wing root and tip. Spent shell casings were jettisoned. Total ammunition capacity was:

| | | |
|---|---|---|
| Fuselage MGs | 2 x 1,000 | = 2,000 rounds |
| Wing MGs | 2 x 500 | = 1,000 rounds |
| | Total | = 3,000 rounds |

**Alignment:** All weapons were normally adjusted to 400 meters direct-fire, which meant that the trajectories of all weapons crossed the parallel sight line (set at 1° 20' against the fuselage axis) at a distance of 400 meters.

On 14 May the *E-Stelle Travemünde* (Group E 5) took charge of the V11 for in-depth testing of the new gun wing. According to BFW's schedule, delivery of production aircraft armed with four MG 17s was supposed to begin on 1 August 1937 at the earliest, however, this deadline was not met. Tests with the V11 were in fact completed at the end of June 1937, however, long-term trials continued until the beginning of October. Here the final round figures from the *E-Stelle*'s "S-Reports": 9 to 15 September 1937: continuation of long-term trials, without stoppages, total number of rounds fired to date:

| | |
|---|---|
| Test stand: | 7,678 |
| Air firing: | 70,639 |
| Total: | 78,317 |

23 to 30 September 1937: No trials, aircraft unserviceable due to airframe problems. 30 September to 7 October 1937: Continuation of long-term trials, stoppages involving control stick grip (KG 12 C) and the machine-guns.

Total number of rounds fired:

| | |
|---|---|
| Test stand: | 7,678 |
| Air firing: | 70,639 |
| Total: | 78,317 |

After this date the V11 no longer appears in the E-Stelle's reports, as it had returned to the factory in Augsburg.

**Bf 109 V12, WerkNr. 1016, D-IVRU**

First flight on 13 March 1937 by Fritz Wendel. Installed power plant was the Jumo 210 D. The aircraft was likewise equipped with the new gun wing; either an MG FF (20-mm) or an MG 17 could be installed on the left side. The right wing was purely a test wing for the MG FF. Testing by BFW was delayed for various reasons, as a result of which the initial firing trial by the *E-Stelle Travemünde* could not take place until 14 September 1937. The V12 was ferried to Travemünde by a BFW pilot (following a reminder from the *E-Stelle* on 27 September 1937). There followed the usual acceptance checks and workshop work in which new production MG FFs were installed.

The very first test-stand firing on 14 October 1937 resulted in loosening of rows of rivets in the shaped attachment section of the forward cannon mount on ribs and secondary ribs. This was traced back to poor work on the part of BFW. Repairs were made by the *E-Stelle*, after which test-stand firing was resumed under closer observation.

There followed twelve gunnery flights (shallow dive), with a progressive increase in rounds fired from 1 to 60. No weapons stoppages. Air firing flights were resumed on 28 October (high and low altitude).

| Total rounds fired at the *E-Stelle*: | | |
|---|---|---|
| Test stand: | MG 17 | 164 |
| | MG FF | 287 |
| Air | | none |

Following the twelfth air firing flight, loose rivets were discovered at the attachment points of the forward cannon mounts in both wings, and BFW was immediately called in. Repairs at the *E-Stelle* began on 19 October; redesign of the involved attachment for production aircraft was now unavoidable.

| Total rounds fired at the *E-Stelle*: | | |
|---|---|---|
| Test stand: | MG 17 | 164 |
| | MG FF | 423 |
| Air | MG 17 | 0 |
| | MG FF | 488 |

Because of the special urgency attached to the project, the loose rivets were replaced with screws. Six air firing flights with strong acceleration about the vertical axis (yaw) and knife-edge flights.

| Total rounds fired at the *E-Stelle*: | | |
|---|---|---|
| Test stand: | MG 17 | 164 |
| | MG FF | 525 |
| Air: | MG 17 | 0 |
| | MG FF | 488 |

After the twenty-first air-firing flight with a total of 1,999 rounds fired in the air, on 13 November torsion wrinkles were discovered on the left side of the combined wing after firing under accelerations about the yaw axis (4 to 5 g). Slight deformations were also discovered on the right wing. The gunnery flights were subsequently abandoned, after which the V12 was transported by rail to BFW, where it was repaired and modified back to production standard (two wing-mounted MG 17s) for further trials.

| Total rounds fired at the *E-Stelle*: | | |
|---|---|---|
| Test stand: | MG 17 | 164 |
| | MG FF | 555 |
| Air: | MG 17 | none |
| | MG FF | 992 |

Based on the testing results, the C- and D-series were equipped with a gun wing mounting two MG 17 machine-guns. Not until May 1938, following various design changes to the wing and test flights by new prototypes, was the 20-mm wing armament cleared for installation in the Bf 109 E-3 series.

**Bf 109 V13, WerkNr. 1050, D-IPKY**

First flight by Dr. Würster on 10 July 1937. This machine was prepared for two special purposes:

- For participation in the VI International Air Meeting in Zurich and
- for a world speed record flight.

The airframe was a standard B-model (without armament), however, a special engine from Daimler Benz was installed. It was a "racing engine" derived from the DB 601 A (DB 601 1/III) with the pre-production series number 160. It was delivered to Augsburg for flight testing on 14 June 1937 and performed satisfactorily in the V13. In July the engine was returned to Daimler Benz, where it was dismantled. Various parts were replaced, and the acceptance run took place on 17 July 1937 and the follow-up run on 18 July. Maximum output was 1,658 H.P. at 2,662 rpm, b = 364 g/PSh and 1.62 atmospheres of boost pressure. The engine was equipped with a VDM three-blade, variable-pitch propeller. After its re-installation and several test flights the V13 was ferried to Zurich (see chapter titled "Air Meeting").

On 29 July the V13 returned to Augsburg with Dr. Würster at the controls, and it was subsequently converted into record configuration in the prototype shop.

The Bf 109 V13 powered by a DB 601 A prior to taking off on a competition flight during the Zurich Air Meet (above). Bf 109 V14 powered by the DB Rennmotor II, which was made available for the Zurich Air Meet.

## Bf 109 V14, WerkNr. 1029, D-ISLU

This aircraft was also earmarked for participation in the Zurich air meet. It was first flown by Dr. Würster on 28 April 1937. Its airframe was also equivalent to B-standard, and at the meeting it was supposed to be flown exclusively by *Generalmajor* Udet. In contrast to the other four participating Bf 109s, which were painted in BFW's typical gray finish for prototypes, the V14 was painted wine-red.

The V14 was powered by the DB 601 *Rennmotor II* racing engine with the pre-production serial number 161. Its peak output was 1,565 H.P. at 2,620 rpm, b = 299 g/

Leichte Wartungsarbeiten an der Bf 109 V13 beim Züricher Flugmeeting.

PSh and 1.62 atmospheres boost pressure. The two racing engines were officially described to the press as 960 H.P. DB 600s.

Engine No. 161 was denied success in Zurich, however. One note: the V14 was flown exclusively by Dr. Würster at Zurich.

Before the participating Bf 109s were flown to Zurich, they were inspected in Augsburg by Obstlt. Werner Junck and Fl.-Haupting. Spies on 14 and 15 July 1937. The occasion was also used to discuss general questions concerning development and improvement of the Bf 109, such as de-icing, design improvements to reduce construction costs, more streamlined mass and aerodynamic balances aimed at reducing parachute hang-ups on bail out, taller radio antenna on the vertical fin for greater radio range, and larger fuel tank for the Jumo 210 and DB 601. Also discussed were improvements to the cockpit: improved view and comfort, panel de-icing, rain-proofing, visibility in rain (bad-weather panel).

Bf 109 C prior to delivery (above left).

Two of the Bf 109 E-4s delivered to Bulgaria (left).

One of the Bf 109 E-3s delivered to Switzerland without armament.

Bf 109 D-1s serving with a fighter training school in 1939-40.

18

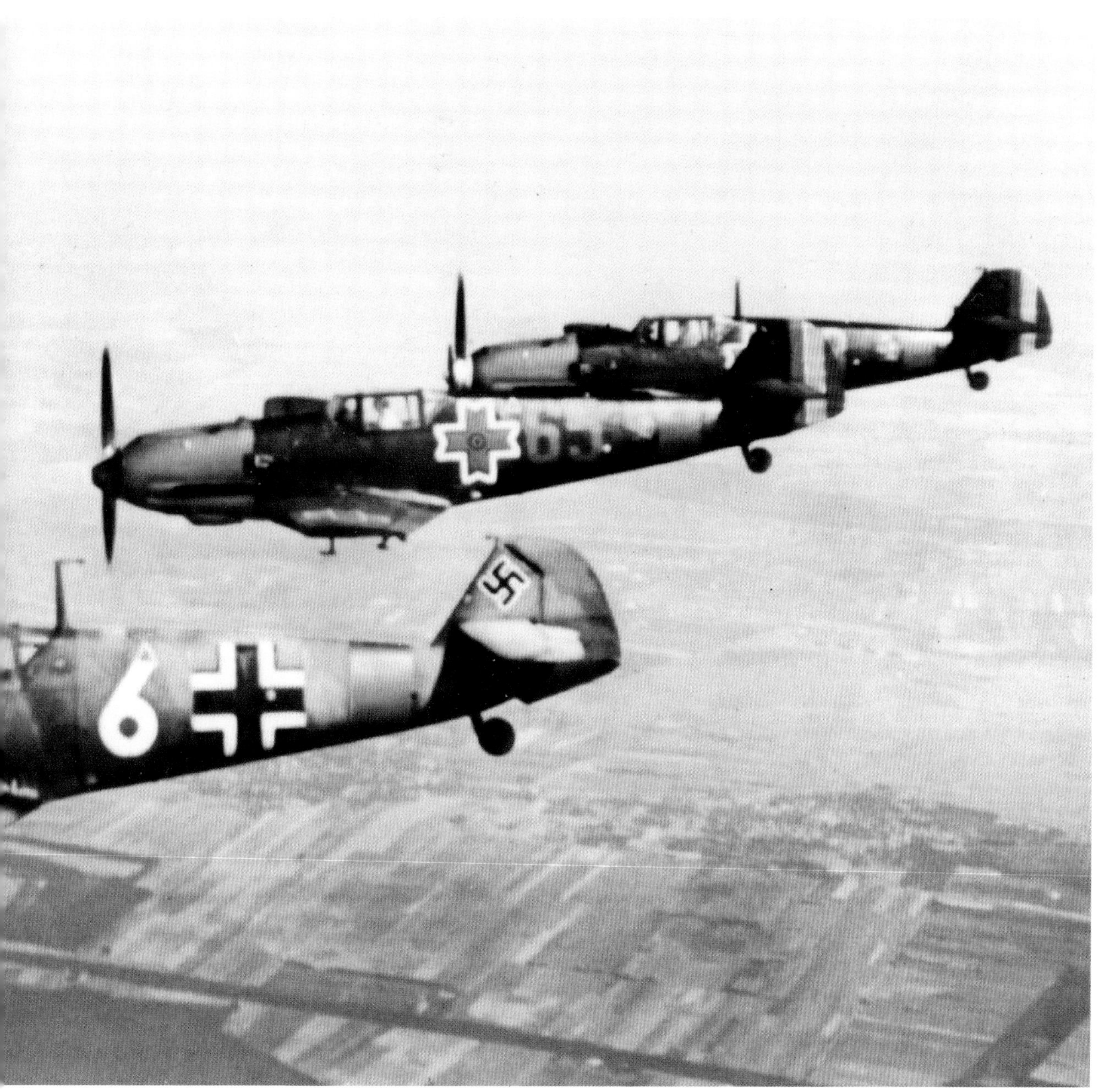

Several Bf 109 E-3s of III./JG 3 during the Battle of France in 1940 (above left).

Bf 109 E-4 of III/JG 53 on a front-line airfield in France (far left).

An unusual formation consisting of one Bf 109 E-4 of the *Luftwaffe* and two Bf 109 E-3a fighters which were delivered to Romania (above).

Captured Bf 109 E-3 at Boscombe Down in England in May 1940 (left).

Unarmed Bf 109 E-0 from the pre-production series serving with the *Jagdflieger-schule Werneuchen* (left).

Bf 109 E-4 in service with the *Luftwaffe*, equipped with two wing-mounted MG FF cannon (above).

Bf 109 E-3 on a forward airfield in 1940 (above left).

Bf 109 E-7 on a frontline airfield in Africa between missions (left center).

*Oberleutnant* Werner Schroer in his Bf 109 E-7/U2 "Black 8" over the Libyan coast (above and left).

# The Zurich Air Meeting

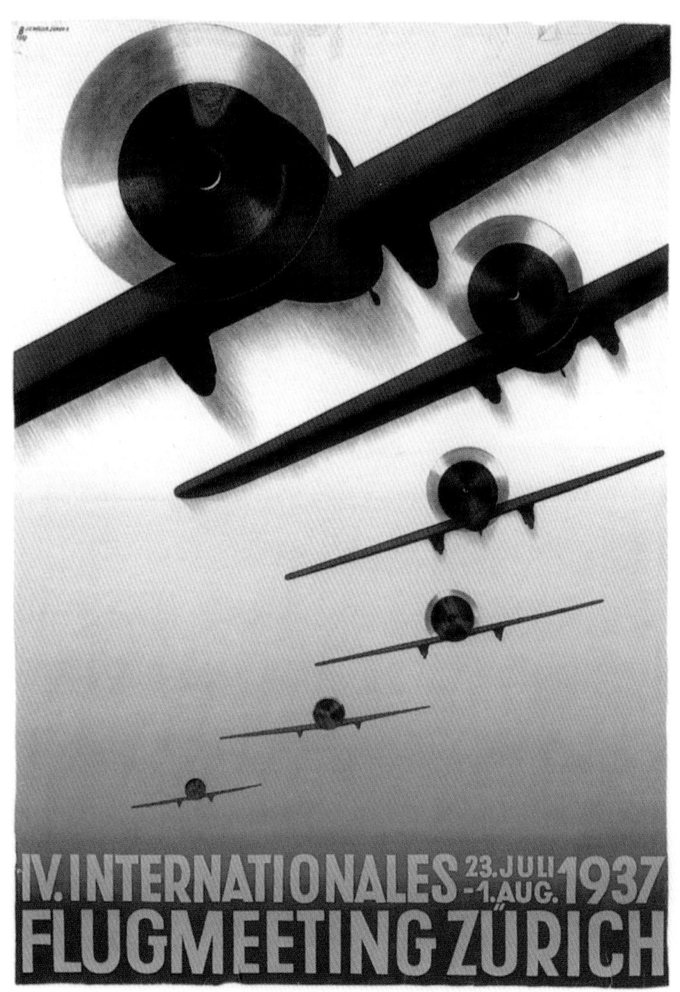


**VI International Air Meeting at Zurich/ Dübendorf from 23 July to 1 August 1937**

Without going into the political motivations which resulted in the extremely strong German participation (40 of 152 entries), there now follows a brief description of the individual competitions in which the participating Bf 109s performed magnificently. In addition to various civil aircraft, for the first time Germany also displayed publicly a number of still little known military aircraft, such as the Do 17 M V1, Hs 123 V5, He 112 A-03 and five Bf 109s:

| | | | |
|---|---|---|---|
| **Bf 109 V7** | D-IJHA | Jumo 210 G | 730 H.P. |
| **Bf 109 V9** | D-IPLU | Jumo 210 G | 730 H.P. |
| **Bf 109 V13** | D-IPKY | DB 601 Racing Engine III, Serial No. 160 | 1,658 H.P. |
| **Bf 109 V14** | D-ISLU | DB 601 Racing Engine II, Serial No. 161 | 1,565 H.P. |
| **Bf 109 B-1/1062** | none | Jumo 210 G | 730 H.P. |

**1 Speed Competitions**
(25 July 1937)

The entire 202-kilometer course was a "rectangular course" which had to be flown four times. Withdrawal of the French entries left only Udet in Bf 109 V14 D-ISLU and the Englishman Gardner in the Per-

cival Mew Gull. It was agreed that Dipl.-Ing. Carl Francke (*E-Stelle Rechlin*) might also participate in the Bf 109 V7 D-IJHA.

Udet suffered engine trouble in the very first lap (engine No. 161, throttle valve lever) and was forced to withdraw. Francke flew the rest of the race in 29 minutes 35.2 seconds, which represented an average speed of 409.64 kph, and emerged the victor.

**The Bf 109 V7 with two different competition numbers (1 and 6) at the Zurich Air Meet (above and top of page).**
**Three Messerschmitts prior to formation takeoff for the Alpine circuit, from left to right Bf 109 B-1, V7 and V9. In the background is the Dornier Do 17 M V1 (left).**

Ernst Udet studies a map in front of the Bf 109 V14; on the right is *Fliegerhauptstaatsingenieur* Roluf Lucht, left Dr. Hermann Würster.

**2 Alpine Circuit for Military Aircraft**
1 July 1937)

This competition consisted of a speed flight over the course Dübendorf—Thun (105 km)—Bellinzona (125 km)—Dübendorf (137 km, total distance 367 km). Times were recorded from takeoff to crossing the target stripe on the specified intermediate airfields and the destination Dübendorf. The circuit was flown in three categories: A = Single-seaters (competition number 6), B = Multi-place (competition number 6) and C = in closed formation of three (competition number 4).

Category A: The circuit was flown in favorable weather. Alongside one French and four Czechoslovakian aircraft, Germany was represented by Udet in the V14 D-ISLU and *Major* Seidemann in the V9 D-IPLU.

During the first stage engine trouble (engine No. 161 again, fractured oil line) forced Udet to make a wheels-up landing near Thun airfield, in which the propeller struck the overhead wires of an electric railway. The fuselage bent almost at right angles behind the cockpit. Udet escaped with minor injuries to one arm. *Major* Seidemann won the race, finishing ahead of the Czech Hlado with a time of 56 minutes 47.1 seconds.

Category B: The winner was the Do 17 M V1, WerkNr. 691, later D-AELE, crewed by:

| | |
|---|---|
| *Major* Polte | pilot |
| G.d.F. Milch | commander |
| Fl.-Ing. Hänsgen | radio-operator |
| Franz | flight engineer |

The Do 17 M V1 was also equipped with special DB 601 engines (Nos. 133 and 162, 1,218 H.P.), which were likewise identified as DB 600 carburetor engines.

Class C: Alpine circuit in formation of three (29 July 1937). The German flight consisted of:

Bf 109 V7, V9 and B-1 (no registration) with pilots Oblt. Trautloft and Oblt. Schleif. The Bf 109 *Kette* won with a time of 58 minutes 52.9 seconds.

**3 International Climb and Dive Competition** (25 July 1937)

Conditions: From takeoff the competitors had to reach an altitude of 4,000 meters above airfield elevation (Dübendorf 440 meters). This was followed by the diving portion of the flight. The dive angles achieved by the German participants, Dipl.-Ing. Carl Francke in Bf 109 V13 and Dipl.-Ing. Schürfeld in Hs 123 V5 were approximately 70 to 75°. The target stripe had to be overflown in a specified direction at a height of at least 100 meters (400 meters maximum). The winner was Francke in the Bf 109 V13 with a time of 2 minutes 5.7 seconds; Schürfeld came second in the Hs 123 with a time of 2 minutes 23 seconds.

Overall, the results were a considerable success for the Bf 109. The question remained open, why the British did not participate with their Hurricane and Spitfire.

**The victors in the Alpine circuit in formation of three, from left to right: Oblt. Fritz Schleif, Hptm. Werner Restemeier, Oblt. Hannes Trautloft (top of page). Major Hans Seidemann (in uniform) in conversation with Dipl.Ing. Carl Francke (above). Ernst Udet (2nd from right), *Major* Carl-August von Schoenebeck (4th from right), Willi Stör (2nd from left), who visited the Zurich Air Meet in an M35, and the aviatrix Liesel Bach.**

Ernst Udet in the Bf 109 V14 (above left).

The Bf 109 V14 was equipped with the DB 601 Racing Motor II (above).

A cracked fuel line forced Udet to crash-land the Bf 109 V14 near Thun, resulting in the fuselage bending almost at right angles (left).

The Bf 109 V14 before takeoff (above center).

During the forced landing near Thun the Bf 109 V14's propeller caught a 15 kV overhead railway power line. Udet escaped with minor injuries (top of page and above).

# The World Speed Record

In 1937 the Class C (land aircraft) world speed record over a three-kilometer course was held by Howard Hughes (567.115 kph). According to the rules of the FAI (*Fédération Aéronautique Internationale*), recognition of a new record required that the former record mark be exceeded by 8 kph.

The machine that was to attempt to set a new record, the Bf 109 V13, which returned from Dübendorf to Augsburg on 29 July 1937, was now modified into the record version in the prototype shop. The measures taken were primarily aimed at improving the external surfaces of the aircraft, which unlike the Me 209 were not significantly different from those of production aircraft. The record aircraft was outfitted with a more streamlined canopy, a smoother forward fuselage with capped propeller spinner, and improved water and oil coolers. Finally, all seams were taped over, and the entire external surface was puttied and polished to a high gloss. Then the aircraft was painted in the typical overall gray finish of BFW prototypes.

Of particular interest, however, is the power plant. It was a *Rennmotor III* (preproduction number 160) developed from the DB 601 A by Daimler Benz (see table below left).

Concerning the record flight itself, the words of Dr. Würster himself (condensed):

"In accordance with the rules of the FAI, the aircraft had to be flown over the three-kilometer course four times (twice out and twice back) at a height of less than 75 meters above the ground. In the vicinity of Bobingen near Augsburg the course followed a perfectly straight railway line. The warning and main signals were indicated by a whitewashed mark. Since the (low altitude) motor developed the most power near the ground, I intended to fly at a height of 35 meters.

Of course, such a high-speed flight requires visibility, no wind if possible, and a favorable sun position. Furthermore, it is necessary for one to be completely familiar with the aircraft beforehand, in order to be able to control the machine at high speed and low altitude with the minimum of control movements. Furthermore, it is necessary to select the best setting for the variable-pitch propeller to provide optimum performance, as well as properly trim the aircraft.

The high speed at low altitude makes it necessary to always train one's eyes three to four kilometers ahead of the aircraft, which means that on passing over the starting point the pilot does not see the end of

| | | | |
|---|---|---|---|
| **Type:** | 12 cylinder, liquid-cooled, fuel-injected motor | | |
| **Configuration:** | In-line engine of vee-configuration with hanging cylinders and a total volume of 33.9 liters | | |
| **Outputs:** | | | |
| Max. output | at 2,650 rpm | at zero altitude | 1,660 H.P. (5') |
| Max. output | at 2,600 rpm | at zero altitude | 1,660 H.P. (10') |
| Short-duration output | at 2,500 rpm | at zero altitude | 1,520 H.P. (20') |
| Short-duration output | at 2,500 rpm | at 500 meters | 1,480 H.P. (20') |
| Continuous output | at 2,400 rpm | at zero altitude | 1,370 H.P. |
| Continuous output | at 2,300 rpm | at 500 meters | 1,120 H.P. |
| Fuel consumption during short-term output | | | 226 g/PSh |
| Fuel consumption at continuous output | | | 220 g/PSh |
| Clockwise-rotating VDM three-blade variable-pitch propeller 2.8 m diameter | | | |
| **Reduction:** | 1 : 1.55 | | |
| **Fuel:** | Mixture 50 percent<br>Special fuel + 50 percent octane<br>100 + 1 percent lubricant | | |
| **Lubricant:** | Racing Oil E "Special" | | |

After returning from the Zurich Air Meeting the Bf 109 V13 was reequipped for an attempt at the world speed record. It was equipped with a DB 601 Racing Motor III, which had a maximum output of 1,660 H.P. The fuselage and wings were puttied and polished to a high gloss to reduce drag.

the three-kilometer course, instead his eyes are already trained one kilometer beyond it. Consequently, during the approach, which was 10 kilometers long, the pilot must 'aim' the aircraft precisely at the course.

In order to get an idea of the weather situation and visibility, I flew the course beforehand in a Bf 108 (D-IGQB). The cloud height was approximately 500 meters. Visibility was adequate, although the sun was already rather low, which was a minor hindrance. There was also almost no wind, so I decided to make the flight and had them prepare the record machine for takeoff. Three observation aircraft (Bf 108s carrying official observers) took off first, one each for the north and south turning points and one to observe the course, and climbed to approximately 250 meters. I needed 17 seconds to fly the calibrated distance each way, which equated to a speed of 170 meters per second.

The engine ran like clockwork during the entire record flight. The turns were high-radius turns, since the machine could not be pulled up in the turn to prevent 'scrounging' speed in the subsequent descent toward the course. Once while turning I flew into a rainbow, but I was also able to overcome this difficulty. So after flying the course four times, at 2 PM on 11 November 1937 (precise times according to the log book: 2:09 25' to 2:31 12' (22 min. 47 sec.) I won the world speed record in the C Class for Germany for the first time with a speed of 610.950 kph.

A more streamlined propeller spinner was also fitted (above)

A word about how the speed was measured: obviously these high speeds could no longer be measured by hand. The FAI also had at its disposal photographic timekeeper devices which were accurate to 1/100 of a second. This was only possible by using special cameras. For measurement

In an attempt to reduce drag the Bf 109 V13 was fitted with a so-called "racing canopy."

they used an Olympia camera (Olympiade 1936) with 50 frames per second and an exposure time of 1/1000 of a second per frame. This camera was developed by the Reich Physical-Technical Institute (PTR). The measuring process involved positioning cameras at the start and finish at a distance of 300 m, which simultaneously photographed the clockwork, the measuring sticks and the time indication on the clockwork."

This ends Dr. Würster's account of his record-setting flight.

The new world speed record of 610.950 kph was recognized by the FAI. It exceeded the old record held by the American Howard Hughes by 44.83 kph.

On the German side, for reasons of secrecy two pieces of false information were given for the entry in the "*Diplôme de record*":

- The aircraft type was identified as the Bf 113 R, and the
- power plant as the DB 600 producing 950 H.P.

But on with the V13: Dr. Würster made eight further record attempts with it between 11 December 1937 and 19 January 1938, however, he failed to break the world record.

In early 1938 the V13 returned to the prototype shop. There it was fitted with a new type of surface evaporation cooling. On 19 April Dr. Würster tested this unusual form of engine cooling as part of the preliminary tests for the Me 209. On 20 July

| Lfd. Nr. des Fluges | Zulassungs-Nr. des Flugzeugs | Führer | Fluggast | Zweck des Fluges | Abflug Ort | Abflug Tag 1937 |
|---|---|---|---|---|---|---|
| a | b | c | d | e | f | g |
| 995 | Bf 108 | Dr. Würster | v. Bethmann-Hollweg | Werkflug | Augsburg | 11/11 |
| 6 | " | " | " | " | " | " |
| 7 | Bf 109 | " | – | Weltrekordflug | " | " |
| 8 | " | " | – | Fotoflug | " | 16/11 |
| 9 | " | " | – | Fotoflug | " | " |
| 1000 | " | " | – | Werkflug | " | 19/11 |
| 1 | Bf 162 | " | – | Versuchsflug | " | 20/11 |
| 2 | " | " | – | " | " | 22/11 |
| 3 | " | " | Wiedemann, Stadler | " | " | 26/11 |
| 4 | " | " | " | " | " | 26/11 |
| 5 | " | " | " | " | " | " |
| 6 | " | " | " | " | " | " |
| 7 | " | " | " | " | " | " |
| 008 | " | " | " | " | " | " |

| Landung | | | geflog. Zeit | | Auftrag | | Bemerkungen 21 |
|---|---|---|---|---|---|---|---|
| Ort | Tag | Uhrzeit | h | min | ~~Auftraggeber~~ | ~~Auftrag Nr.~~ | |
| i | k | l | m | n | Werk-Nr. | D | q |
| | | | 502 | 00 | | | |
| Augsburg | 11/11 | 8 45 | | 28 | 1568 | 16 QB | |
| " | " | 13 12 | | 12 | 1568 | " | 610,95 km/h |
| " | " | 14h31'12" | | 2'14,7" | 1050 | IPKY | Weltrekord für Landflugzeuge |
| " | 16/11 | 14 02 | | 11 | 881 | IJHA | |
| " | " | 14 26 | | 9 | " | " | |
| " | 19/11 | 11 49 | | 13 | 884 | IXZA | |
| " | 20/11 | 15 46 | 1 | 14 | 818 | ADBE | |
| " | 22/11 | 12 24 | | 11 | " | " | Kunstflugvorführ. vor Adolf Hitler |
| " | 26/11 | 13 54 | | 5 | " | " | |
| " | " | 14 02 | | 5 | " | " | |
| " | " | 14 13 | | 8 | " | " | |
| " | " | 14 20 | | 4 | " | " | |
| " | " | 15 44 | | 7 | " | " | |
| " | " | 15 51 | | 4 | " | " | |
| | | | 505. | 32 | | | |

Dr. Hermann Würster in the cockpit of the Bf 109 V13 prior to the record-breaking flight (above).
A copy of the page from Dr. Würster's log book containing the entry for the record-breaking flight of 11 November 1937 (left).

Right: The official certificate from the FAI for the world record flight with a speed of 610.950 kph. It contains two pieces of false information, released for propaganda purposes: the Bf 109 V13 was designated the Bf 113R, and instead of the DB 601 R III the engine was given a DB 600.

Below: Caricature by Ernst Udet poking fun at the competition for the world record between Willy Messerschmitt and Ernst Heinkel. Just seven month's after the Bf 109 V13's record-setting flight, on 5 June 1938, Ernst Heinkel's He 112U, actually the He 100 V2, set a new speed record of 634.473 kph.

**fédération aéronautique internationale**

siège social : 6, rue galilée, paris

# diplôme de record

**nous soussignés certifions que** le Dr. Ing. Hermann Würster,

sur monoplan Bf 113 R., moteur DB 600-950 PC, 12 cyl., **a établi le** 11 Novembre 1937 **le**

**record suivant :** Vitesse sur base (610.950 km/h.)

à Augsburg.

Classe C.

le président de la f. a. i. :

pour l'Aero-Club von Deutschland,

le président :

le secrétaire général :

coquemer, grav.

„Der Stein des Anstosses“

he made a successful forced landing at Lechfeld after the machine ran out of coolant. The machine was undamaged, and on the same day Würster flew it back to Augsburg. Two more flights were made by him on 21 July. Further surface evaporation cooling flights were now undertaken by Fritz Wendel, last entry: 5 September 1938. On 30 November 1939 Wendel flew the aircraft to the *Flieger-Technischen-Schule* (FTS) in Munich. From there the trail of the record-setting Bf 109 V13 D-IPKY disappears.

# From the Bf 109 C/D to the Bf 109 V15

### Bf 109 C- and D-Series

The Bf 109 V11 and V12, which were equipped with the new "gun wings," may be seen as the prototypes for these series. The C-1 was the only version of the C-series to be built, and all were constructed in the Augsburg factory. A total of 58 left the assembly halls.

The C-3 version, which appeared in the "Type Lists" on 26 August 1940, was planned only as a precaution in the event that the wing-mounted MG FF cannon became a possibility before the start of E-series production. It remained in the planning stage.

### Export

The RLM was very interested in selling aircraft released for export, other aviation-related equipment and manufacturing licenses on account of Germany's generally poor foreign currency situation. In August 1938 the Messerschmitt company began negotiations with the Japanese government for the license production of the Bf 109 in Japan. The value of the license was approximately 700,000 Reichsmark. Japanese interest in the Bf 109 waned noticeably, however, as they hoped to produce the He 100 under license. Japan first received two examples of the aircraft in 1940.

**Ready for collection: Bf 109 D with metal variable-pitch propeller and oil cooler beneath the left wing.**

**Aircraft Description**

General
The Bf 109 C and Bf 109 D are single-seat, single-engine light fighter aircraft.
The Bf 109 C is equipped with a Jumo 210 G (fuel-injected engine) and the Bf 109 D with a Jumo 210 D (carburetor engine).

| | |
|---|---|
| **Design Configuration** | All-metal cantilever monoplane |
| Retractable undercarriage | |
| Slats and landing flaps | |
| Enclosed cockpit | |
| **Structural Strength** | The aircraft type is designed for |
| | Usage group H |
| | Stress group 5 |
| Maximum takeoff weight | 2 160 kg |
| Maximum horizontal speed At ground level | 410 kph |
| Maximum allowable diving speed | 800 kph |
| Maximum allowable speed with landing flaps fully extended | 250 kph |
| Aerobatic suitability | the aircraft is fully aerobatic |
| **Airframe** | See description of the Bf 109 B series |

**Dimensions**

| | |
|---|---|
| Wingspan | 9.9 m |
| Overall length | 8.7 m |
| Maximum height tailwheel down | 2.5 m |
| Wheel base | 2.0 m |
| Aerodynamic surface | 16.40 m$^2$ |

| Characteristic Values[8] | | Bf 109 C | Bf 109 D |
|---|---|---|---|
| Wing loading | kg/m$^2$ | 131.7 | 131.72 |
| Weight-power ratio | kg/H.P. | 2.88 | 3.13 |
| Wing performance | H.P./m$^2$ | 45.7 | 42.2 |

| Climb performance | | Bf 109 C Jumo 210 G | Bf 109 D Jumo 210 D |
|---|---|---|---|
| Maximum allowable takeoff weight with VDM propeller | kg | 2 160 | 2 160 |
| Maximum speed at ground level | kph | 436 | 435 |
| Maximum speed at maximum boost altitude (5.25 and 4.0 km respectively) | kph | 498 | 471 |
| Time to climb to 5 000 m | min | 7.13 | 8.51 |
| Service ceiling | m | 9 000 | 8 100 |
| Endurance in full throttle flight at maximum boost altitude (5.25 and 4.0 km respectively) | hrs | 2 | 2 |
| Landing speed | kph | 111 | 111 |

[1] Characteristic values. Values for both types regarding high-speed flight are based on 5 min short-duration output at maximum boost altitude and a takeoff weight of 2 160 kg. All values take into consideration ram effect except in the climb.

**Performance Sheet Bf 109 C**
Range, climb and speed performance

| Height performance m | Max speed rate of climb kph | Climb min | Forward speed at best kph |
|---|---|---|---|
| 0 | 410 | 0 | 216 |
| 1 000 | 430 | 1 | 224 |
| 2 000 | 440 | 2.3 | 232 |
| 3 000 | 445 | 3.6 | 242 |
| 4 000 | 460 | 5.4 | 250 |
| 4 500[1] | 465 | | |
| 5 000 | 450 | 7.5 | 258 |
| 6 000 | 430 | 10.3 | 266 |
| 7 000 | 400 | 13.4 | 275 |

[1] Maximum boost altitude 4.5 km
Speed values with ram effect
Service ceiling 9 000 m
Range: 1 hr 50 min cruising flight at max boost altitude 4.5 km

**Performance Sheet Bf 109 D**
Range, climb and speed performance

| Height performance m | Max speed rate of climb kph | Climb min | Forward speed at best kph |
|---|---|---|---|
| 0 | 410 | 0 | 224 |
| 1 000 | 430 | 1.35 | 231 |
| 2 000 | 440 | 2.87 | 238 |
| 2 400 | 445 | | |
| 3 300[2] | 450 | | |
| 4 000 | 445 | 6.23 | 251 |
| 5 000 | 430 | 8.51 | 257 |
| 6 000 | 410 | 11.60 | 264 |
| 7 000 | 385 | 16.40 | 270 |

[2] Maximum boost altitude 3.3 km
Speed values with ram effect
Service ceiling 8 100 m
Range: 1 hr 50 min cruising flight at maximum boost altitude 3.3 km
Armament: two synchronized MG 17s in the fuselage and two unsynchronized MG 17s in the wings.
Description see under Bf 109 V11
Production numbers

| | | |
|---|---|---|
| Bf 109 C-1 | BFW* | 58 aircraft total |
| Bf 109 D-1 | BFW* | 4 aircraft total |
| Focke-Wulf | | 123 aircraft total |
| Erla | | 168 aircraft total (guide price RM 85,000) |
| Fieseler | | 80 aircraft total |
| AGO | | 128 aircraft total |
| Arado/W | | 144 aircraft total |
| | Total | 647 aircraft |

(*BFW = Augsburg factory)

The external differences between the C- and D-series are difficult to ascertain.
Comparison with the B series is rather easier. The most obvious differences are:
gun openings in the wings (also D)
always has variable-pitch propeller (also D)
always has antenna mast (also D)
short, somewhat larger exhausts

External identifying features of the Bf 109 D compared to the Bf 109 B/C:
Longer exhaust stacks, in rare cases exhaust arrangement similar to that of the E-series
Canopy with rather more vertical windscreen
Torque link on tailwheel
It should be noted that irregularities in external appearance result from independent modifications and conversions by the units (front-line maintenance facilities), consequently, it is not always easy to identify the C or D in photographs.

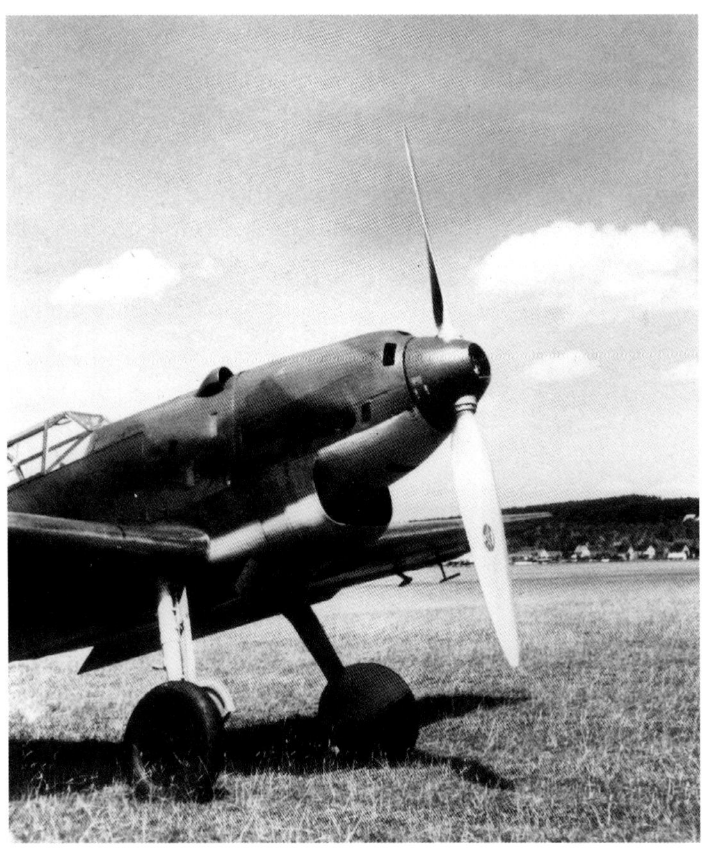

The Bf 109 made such an outstanding impression at the Zurich Flying Meeting that Switzerland ordered ten Bf 109 D-1s (above). The Bf 109 C was armed with four MG 17s (right).

Switzerland was the first foreign nation to receive the Bf 109 for its air force, the *Flugwaffe*. A purchase agreement for five Bf 109 D-1s and one Bücker 180 was concluded in 1938. The first Bf 109 D-1 was delivered in December 1938. The purchase price for the five Bf 109 D-1s and the one Bü 180 was RM 731,000. The contract was increased by five further D-1s, which were delivered during the course of January 1939. The *Flugwaffe* assigned the following serial numbers to the aircraft:

The first Bf 109 E-3s were delivered to Switzerland in May 1939.

| Serial Number | WerkNr. | Delivery Date |
|---|---|---|
| J 301 | 2297 | 19/01/1939 |
| J 302 | 2299 | 19/01/1939 |
| J 303 | 2295 | 17/12/1938 |
| J 304 | 2298 | 10/01/1939 |
| J 305 | 2300 | 05/01/1939 |
| J 306 | 2301 | 10/01/1939 |
| J 307 | 2302 | 07/01/1939 |
| J 308 | 2303 | 10/01/1939 |
| J 309 | 2304 | 19/01/1939 |
| J 310 | 2305 | 05/01/1939[1] |

[1] Shot down in aerial combat with Bf 110s over Boncourt on 4 June 1940.

This export variant received the designation Bf 109 E-3a. The first batch delivered comprised 56 examples. Another 24 followed in 1940, for a total of 80 Bf 109 E-3a fighters.

### The Forerunners of the Bf 109 E-Series

One might actually include the Bf 109 V13 and V14 among the prototypes for the new E-series, because they were powered by the DB 601. These power plants were not production engines, however, instead they

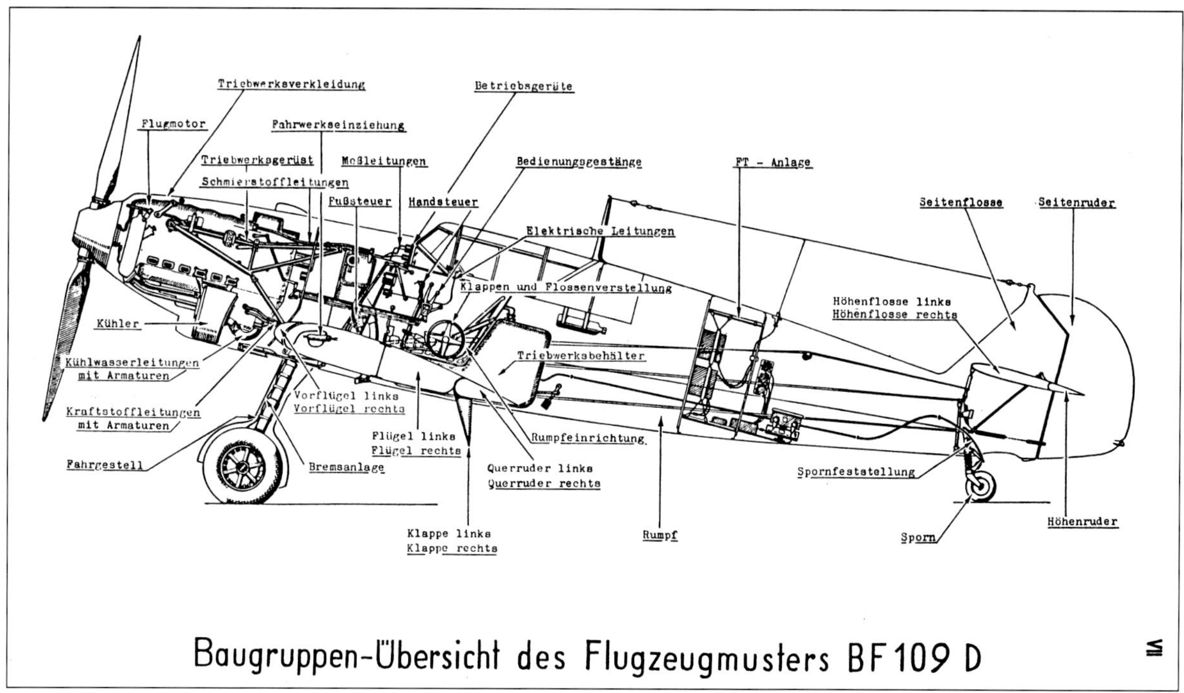


**Bf 109 D component overview. Below: Preflight preparations for Bf 109 Cs of a fighter training school: the inertia starter hand crank was turned by two ground crew ("black boys").**

were special motors. In addition the two unarmed aircraft were used for special purposes (air meet and world record attempts). Thus, the first real prototypes of the E-series were the Bf 109 V15 and V15a.

**Bf 109 V15, WerkNr. 1773, D-IPHR**
(as of January 1940 CE + BF)

The aircraft made its first flight on 18 December 1937 in the hands of Dr. Würster. The aircraft was powered by a DB 601 A fuel-injected engine (serial number 148) with single-stage supercharger (4 km) producing 1,100 H.P. for takeoff. The engine drove a three-blade VDM variable-pitch propeller with a diameter of 3.10 meters. The new engine was approximately forty centimeters longer than the Jumo.

The transition from the Jumo 210 engines used to date in the prototypes and the A-, B-, C- and D-series to the roughly fifty percent more powerful DB 601 meant that various design changes were required in order to accommodate the new motor. These included a redesigned engine cowl-

ing. The water and oil cooling systems were completely redesigned. The water cooling system was divided into two parts: a radiator was located under each wing close to the wing root in front of the landing flap. Each was equipped with an adjustable radiator shutter (manually operated). Cooling took place in a closed cooling circuit. The oil cooler was located beneath the forward fuselage, and the oil tank on the inner side of the fuselage end bulkhead with an external access point.

On each side of the aircraft there were six ejector exhausts, each with upper and lower exhaust shields.

The increase in takeoff weight (of approx. 450 kg) meant that the fuselage structure had to be beefed up, as well as the undercarriage. Because of the DB 601's greater fuel consumption, the capacity of the seat-shaped fuel tank was increased to 400 liters.

The welded steel tube engine mount of the Jumo-powered versions was replaced by two engine bearers made of pressed *Elektron*. These were mounted onto the fuselage attachment points and were braced against the undercarriage gussets by streamlined struts.

A supercharger air intake had to be fitted to the engine cowling in order to provide air to the supercharger. While the extended supercharger air intake used by the *Rennmotoren II* and *III* (Bf 109 V13 and V14) had proved to be less than perfect aerodynamically, it was retained for some time (E-0 series). Extensive trials and testing finally resulted in the most favorable intake shape.

Armament consisted of two MG 17s with EPAG in the fuselage and two MG 17s with EPAD in the wings.

The obligatory 1 1/2 hour flight was completed on 4 February 1938, and the V15 was flown to Rechlin (E 5) the same day. In April, however, it was back in Augsburg for side-by-side trials in May aimed at ironing out a problem that had developed with the V15a. The aircraft remained at Augsburg until January 1940 (code changed to CE + BF), where it was used for various test purposes in 1938 and 1939. One special experiment was the lift spoiler trials in July and September 1939. The flights were carried out by Dr. Würster and Fritz Wendel. The retractable spoilers were located on the upper surface of the wing; they reduced lift and thus decreased landing speed. The trials were part of the preliminary work on the Bf 109 T carrier version. The following remarks concerning the V15 appear in the *C-Amt* program of 1 July 1940:

"From April 1940 as T carrier aircraft at Focke-Wulf in Bin.-Johannisthal." This would indicate that the V15 had been converted into the T-version by this time (extended wing, catapult fittings, arrestor hook, etc) or was in the process of being converted.

### Bf 109 V15a, WerkNr. 1774, D-ITPD

First flight on 21 April 1938 by Dr. Würster, constructed to the same standard as the V15. The test record of 26 April 1938 (flight by Dr. Würster) reveals that the installed DB 601 was producing 200 H.P. less than expected at altitude. Experts from Daimler Benz proposed the following possible reasons:

- Ingestion of exhaust gas by the new supercharger air intake.
- Excessive reduction in fuel pressure at altitude.
- Shifting of ignition timing as a result of drop in oil pressure at altitude.
- Changing of the mixture composition.
- Serious drop in propeller effectiveness at altitude.

All of these points were examined by BFW in flight tests. A parallel test with Bf 109 V15 revealed that this machine's engine also failed to produce the required power at altitude. This test was flown by Dr. Würster on 6, 7 and 9 May, and engine output was off by 146 H.P. at an altitude of 3,900 meters. Two different versions of supercharger air intake were also tested, however, no differences could be ascertained at maximum boost altitude.

In order to eliminate the ingestion of exhaust gas, flights were conducted with two styles of exhaust which reliably prevented exhaust gas from reaching the supercharger air intake:

- Exhausts which projected 150 mm above the exhaust shield.
- Collector exhausts.

Fuel pressure at an altitude of 5 kilometers was 1.75 kg/cm$^2$, while above 5.5 km the indicator wandered all over the scale. Since the ignition timing could switch to "late" when oil pressure fell, a flight was undertaken with the control locked in the "early ignition" position.

During one flight the mixture remained set at "weak." As a result, engine temperature rose on the ground and at altitude, as a result of which the radiators had to be opened further.

Purely theoretically, propeller effectiveness should have been greater at altitude rather than less (based on manufacturer's guarantee and previous experience). During the course of the trials the following were used on the supercharger air intake side: normal ejectors, longer exhausts, exhaust collector, and on the V15 the extended supercharger air intake. There is no information available concerning the solution of the problem. On 7 June 1938 the Bf 109 V15a was flown to the *E-Stelle Rechlin.*

# Bf 109 E-Series

Testing of the two prototypes for the E-series, the V15 and V15a, revealed various shortcomings, which had to be eliminated before the start of quantity production.

The principal source of difficulties was the DB 601 power plant, as there were problems with the water and oil cooling systems. A new, rather larger oil cooler was therefore developed, and this performed satisfactorily in subsequent trials.

New coolant radiators were also installed. By January 1938 the power plant situation was satisfactory enough for the *E-Stelle Rechlin* to begin testing the E-variants. Initial testing took place in Augsburg, however.

Additional tests were carried out with the objective of equipping the first E-series aircraft with a fuselage-mounted armament of two MG 17s and one MG FF. The installation would have been possible if the oil tank had been relocated either to the wing or to the fuselage behind the pilot's seat. The most serious doubts about this solution were related to power plant operation. Consequently, such an installation was abandoned.

Like the V15 and V15a, the first major production variant of the Bf 109, the "E" or "Emil," differed from the A- to D-series mainly in its new, more powerful power plant, which gave the front end of the aircraft an entirely different appearance. The DB 601 (fuel-injected) engine, which was declared fit for service introduction in 1937, brought with it an increase in power of about 300 H.P., which resulted in an increase in maximum speed of just under 100 kph compared to the C- and D-series.

But the new E-series also had its disadvantages. The greater weight of the DB 601 (approximately 170 kg) and extensive accessories resulted in a significant weight increase (of approximately 450 kg), which brought with it a higher wing loading and consequently a higher landing speed, and a greater tendency for the aircraft to ground-loop. This led to an increased number of accidents, most of which resulted in damage to the undercarriage and wings, which the records of the repair industry for the Bf 109 clearly show. The cases of undercarriage damage were also partly due to the in some cases inadequate training of pilots converting to the Bf 109. Effective flying time was reduced as a result of the 33.9-liter engine's greater fuel consumption. Some E-variants made up for this through the use of an external 300-liter drop tank, which increased range. This presupposed

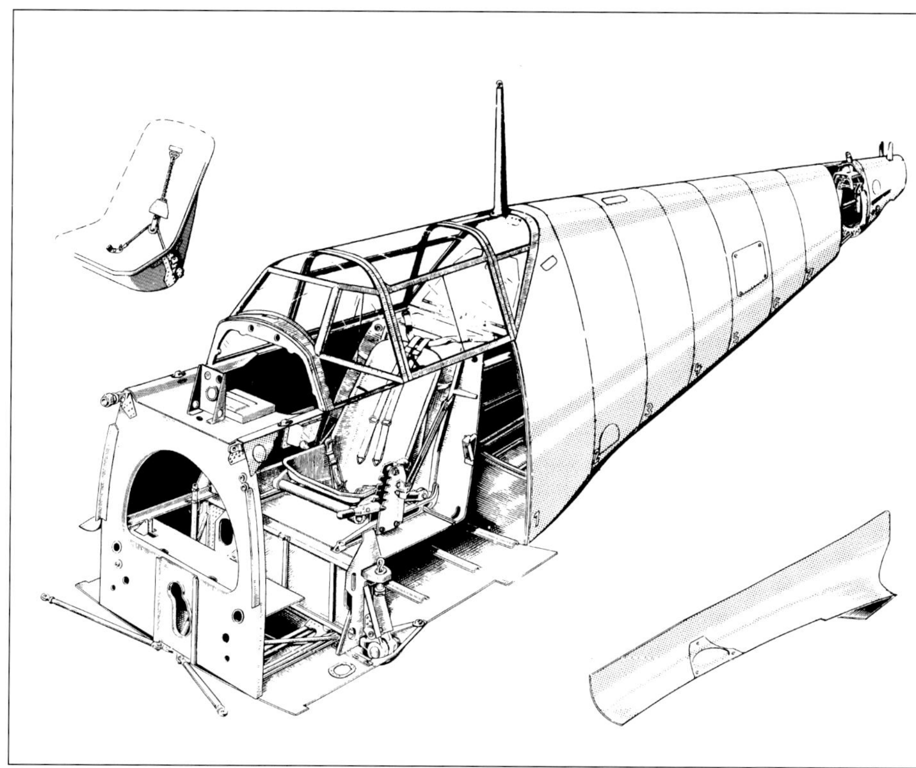

View of a new Bf 109 E-0, the first "Emil" (above). Illustration of the monocoque construction used for the fuselage, shaped Dural sheet sections with attached bulkheads stiffened by longitudinal stringers (left).

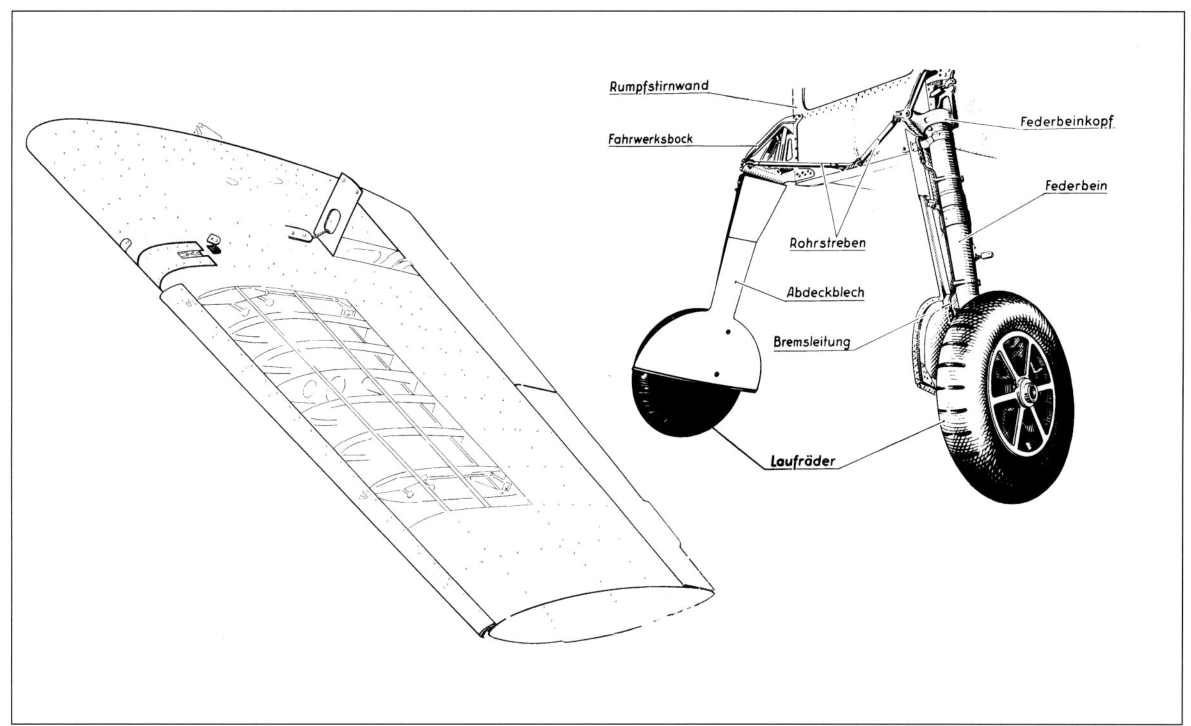


the use of a bomb-dropping system, however, whose mounting was modified for carriage of the tank. In the beginning there was little change in armament, since at the time of the E's introduction there was no reliable 20-mm engine cannon available, and there were also difficulties with the fuselage installation for the MG FF. Consequently, plans to introduce an engine-mounted cannon had to be shelved for the time being. Beginning with the Bf 109-3, however, the aircraft's armament was bolstered through the addition of two MG FF cannon in a modified wing. The MG FF fired standard cannon ammunition, while the later MG FF/M used armor-piercing H.E rounds; both types were far more effective than the 7.9-mm bullets of the MG 17.

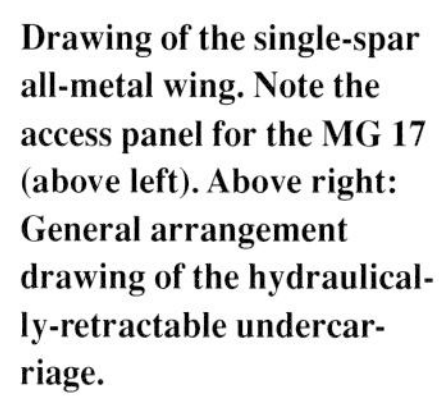

**Drawing of the single-spar all-metal wing. Note the access panel for the MG 17 (above left). Above right: General arrangement drawing of the hydraulically-retractable undercarriage.**
**Bottom: Cutaway drawing of the DB 601 A twelve-cylinder inverted-vee engine. Clearly visible are the supercharger, the crankshaft with balancing weights and the hollow propeller shaft for engine-mounted weapons.**

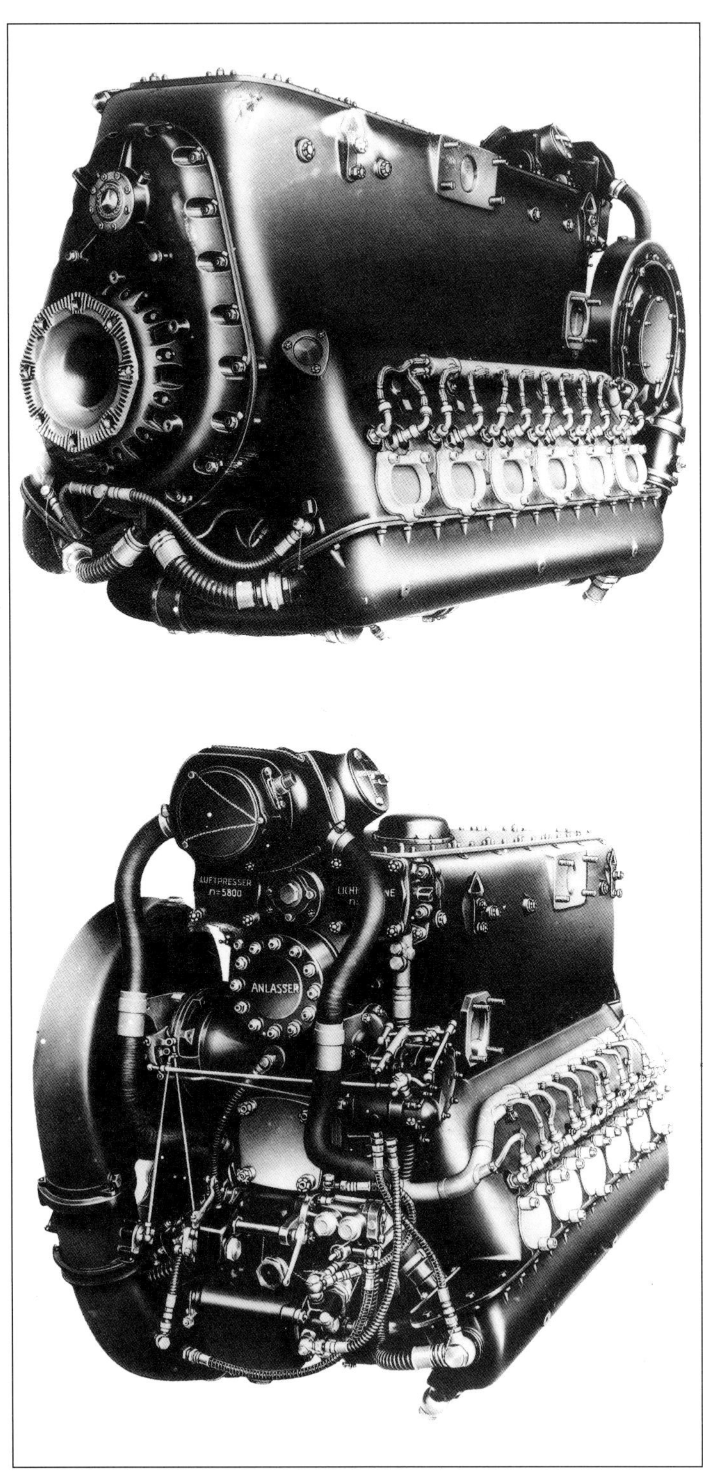


**Two views of the DB 601 A, from the front showing the hollow propeller shaft and from behind the supercharger and starter.**

### Bf 109 E-0

A total of ten examples of the pre-production variant were built in the Augsburg factory, where acceptance flights were carried out in the second half of 1938. The aircraft were powered by the DB 601 A, armament consisted of four MG 17s, two mounted in the fuselage and two in the wings. *Werknummern* assigned to these aircraft were:

| | | | | | |
|---|---|---|---|---|---|
| 1781 | D-IECY | 1784 | 1787 | 1790 | D-IFHG |
| 1782 | | 1785 | 1788 | | |
| 1783 | | 1786 | 1789 | D-INDG | |

### Bf 109 E-1

Two examples of this variant were built in the Augsburg factory and test-flown in November 1938. According to Delivery Plan No. 10 of 1 January 1939, initial plans called for the production of fourteen machines. But after a modification of the delivery date and procurement numbers on 3 February 1939, Augsburg built just *Werknummern* 1791 and 1792. The remaining twelve machines were built by licensees. The E-1 was powered by the DB 601 A. Armament consisted of two synchronized MG 17s in the fuselage and two unsynchronized MG 17s in the wings. For a description of the ammunition feed system see the section on the Bf 109 V11. These aircraft were dedicated test machines which were flown by the Messerschmitt A G in Augsburg as so-called "loan aircraft" (from the RLM).

### Bf 109 E-1, WerkNr. 1791, D-IQCP
(CE + BJ)

The aircraft made its first flight in the hands of Messerschmitt acceptance pilot Helmut Kaden on 16 November 1938. The aircraft remained at Augsburg until August 1941. The following tests were carried out with the aircraft (chronologically):

- Prototype for the Bf 109 E-1, handling quality and stability measurements.
- Special self-alignment tests. Installation for 110% performance, locking tailwheel.
- Installation of head armor, testing of automatic propeller pitch control.

**With its improved armament (2 MG 17 and 2 MG FF) the Bf 109 E-3 quickly replaced older Bf 109 B.**

- Stability measurements with and without wing root fairings (filets).
- Gliding trials (June 1940).
- Lengthened tailwheel (11 November 1940 to 20 February 1941).
- Lippisch slats (March 1941), prototype intake scoop, pilot aiming system.[1]
- As of August 1941 the aircraft was ready for collection by the *E-Stelle Rechlin*.

**Bf 109 E-1, Werk.Nr. 1792, D-IUFG** (CE + BK)

This machine was also a "loan aircraft," which was test flown in November 1938. The precise date and the pilot's name are not known.

The machine was initially used for tests with the automatic pull-out system. Then it served as a prototype for modifica-

[1] The aiming system was the OG 19 A (optical device) made by the Steinheil firm. It was tested on the ground and in the air. Results showed that the complete inability to accurately estimate distances resulted in a danger of ramming. The device also made it extremely difficult to monitor the instruments in flight. The system was rejected (Messerschmitt test report No. 109 11 E41).

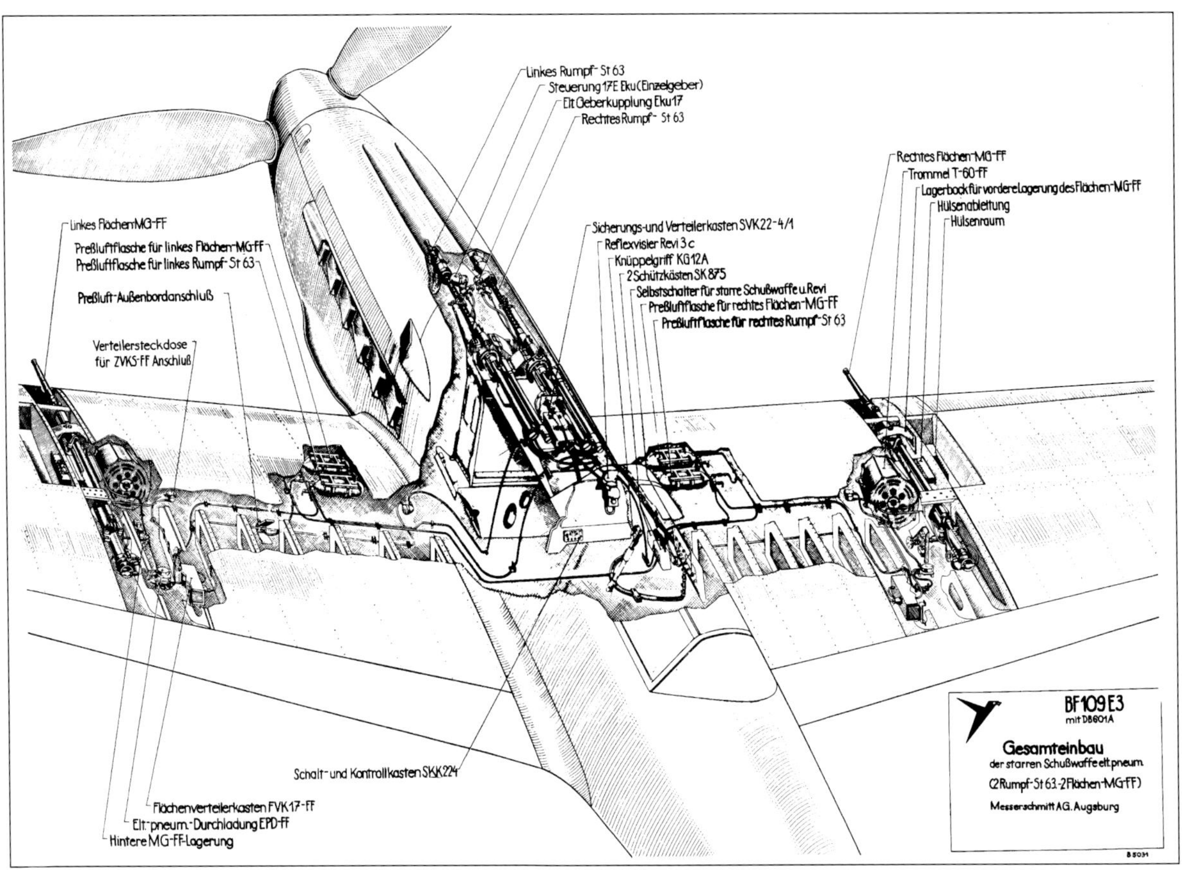


**General arrangement drawing of the guns carried by the Bf 109 E-3.**

tion directives for the E-1. As well, it was used to test the Everling slat. On 9 August 1940 the aircraft was involved in a crash-landing at Augsburg. Since this *Werknummer* appears neither in the subsequent *C-Amt* programs (prototypes), nor in Messerschmitt pilot log books, it may be assumed that the machine was written off.

Brief description from "Type Summary" (GL/C-E No. 384/42 secret):

| | | |
|---|---|---|
| Bf 109 E-1 | Engine: | DB 601 A |
| Fighter | Armament: | 4 MG 17s |
| | | (fuselage and wings) |
| | Fuel: | 400 l |
| | Radio: | FuG VIIa |
| | (with E-1 S inflatable life raft) | |

### Bf 109 E-1/B

The letter "B" stood for *Bombenanlage* (So-3), or bomb installation. This made it possible for the aircraft to be used as a fighter and a dive-bomber. Later on the designation *Jagdbomber* (*Jabo* = fighter-bomber) came into general use. It was also possible to install bomb-dropping equipment on the E-4, E-4/N and E-7, and later on the F-series, as well. The following types of bombs (and later bomb dispensers) were used depending on the nature of the target, however, their use required three different bomb carriers:

1 x SC 500 kg or
1 x SD 500 kg or
1 x SC 250 kg or
1 x SD 250 kg or
4 x SC 50 kg or
96 x SD 2 (2 kg)

### Description of the Carriers:
### 1. ETC 500

An ETC 500 (***E****lektrische* ***T****rägermittel für* ***C****ylinderbomben*) was attached to the underside of the fuselage by means of four bolts and was covered by a two-piece fairing (left and right halves). The bomb (or AB = Bomb Dispenser) was held in position by four adjustable paws. The bomb clasp, which served to release the bomb, was attached to the bomb and locked into the carrier. Like all external stores, the bomb was released electrically by means of a button on the control stick grip. The bomb could also be released by pulling an emergency release handle. The type of sight used was the Revi C 12/d.

Bombing attacks using SC 250 or 500 bombs were carried out in dives of up to 45°. At a dive angle steeper than 65° the bomb struck the propeller within .03 seconds, which posed a serious threat to the pilot's safety! (detonation of the bomb if fuse set to "oV" = no delay).

### 2. Rack with Four ETC 50 VIII

Instead of the ETC 500, a rack with four ETC 50 including fairings was installed. The bombs, hung in pairs one behind the other, could likewise be dropped in diving flight.

### 3. Bomb Carrier for SD 2

The bomb carrier was hung on the fuselage attachment points. Four racks for 24 SD 2s were installed on the carrier and covered with a metal fairing. The mode of attack was low-altitude level bombing (bombs dropped in sequence). The SD 2 close-support bomb had a very great fragmentation effect and was used mainly against living targets (marching columns, troop concentration areas in forests, etc.) or against parked aircraft. Maximum allowable speed with bombs in place was 600 kph.

Loading: the 250- and 500-kg bombs had to be loaded using a hydraulic bomb trolley (LWC 500), while the 50-kg bombs and the SD 2 could be loaded by hand.

### 4. 300-Liter Drop Tank Installation (from E-7/N)

A simple conversion allowed the Bf 109 E/B to be flown with a 300-liter Junkers drop tank in place of the bomb-carrying equipment. The carrier was attached with the fairing and contained the mounting clasp on which the drop tank hung by a steel strap. Four adjustable support struts stabilized the tank during flight. When an aircraft was equipped in this way an additional 9-liter oil tank had to be installed on the left side of the engine to compensate for the increased oil consumption.

While in use, air from the supercharger (reduced to 0.34 atmospheres) forced the fuel from the drop tank through the circulation and filling lines into the main tank. While carrying the drop tank the aircraft was no longer fully aerobatic, and it had to be jettisoned as soon as contact was made with the enemy.

The variants so equipped were the Bf 109 E-7 (2 x MG 17 and 2 x MG FF) and E-8 (4 x MG 17).

110 examples of the Bf 109 E-1/B were built, 61 by the Fieseler Werken (GFW) and 49 by Arado in its Warnemünde factory (ArW). The following brief description is from the "Type Summary":

### Bf 109 E-1/B

Dive-bomber of fighter bomb-dropping system with So-3 installation.

## Bf 109 Test-bed for Air-dropped Weapons

The continuous changes and improvements made to the ETC 500 and the drop tank jettisoning system made various flight tests necessary. The following are two examples:

The first test-bed for the ETC 500 bomb-dropping system was Bf 109 E-3 WerkNr. 1361, CA + NK. The trials began in October 1939. Then on 26 June 1940 the aircraft was flown from Rechlin to Augsburg for installation and testing of the new ETC 500 XIb. At the same time tests were being conducted with the aim of increasing the aircraft's range. On 3 August 1940 the machine returned to Rechlin, where on 10 August it was involved in a belly landing. The damage was not repaired.

**From October 1939 the Bf 109 V26 was used as a test-bed for the ETC 500 external stores rack (right), later to test the SD 250 fragmentation bomb with circular tail fin (below). Bottom of page: Graph depicting the Bf 109 E's maximum speed at various altitudes.**

As a replacement, Messerschmitt-Augsburg modified Bf 109 E-4/N WerkNr. 3744, CI + EJ, for long-range operation. On 3 September 1940 Messerschmitt test pilot Karl Baur ferried it to Rechlin for further testing.

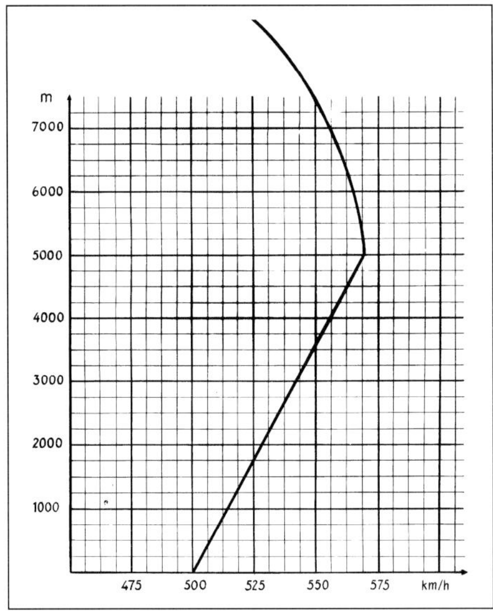

## Bf 109 E-2

This variant did not enter production. In many publications the Bf 109 E-3 WerkNr. 1952, CE + BM, is identified as the "prototype Bf 109 E-2." However, the object seen projecting from the aircraft's spinner in the familiar photos is not the muzzle of an engine-mounted cannon, but instead a tow coupling!

Tests with the Bf 109 V26 and the SC 500 bomb revealed that ground clearance was inadequate (top). As a result a change was made to the smaller SC 250, here seen being loaded with the help of a LWC 500 hydraulic bomb trolley.

## Bf 109 E-3

This type was the most-produced of all the E-variants and became the *Luftwaffe*'s standard fighter aircraft in 1940-41. It was not completely replaced until the introduction of the improved Bf 109 F in 1941. The airframe was similar to that of the Bf 109 E-1 with only minor structural strengthening. The DB 601 A power plant was retained. The most effective innovation, however, was the standard installation of two MG FF cannon in the wings, which represented a significant increase in firepower. Each gun was fed by a T 60 FF ammunition drum holding sixty rounds; these were installed from below and were covered by bulged fairings.

### Installation

Two versions of the MG FF were used, the "A" and "B." The "A" had an electrical-pneumatic firing system (EPA-FF), while the "B" was fired electrically (EA-FF). Both types came with an electrical-pneumatic cocking system (EPD-FF).

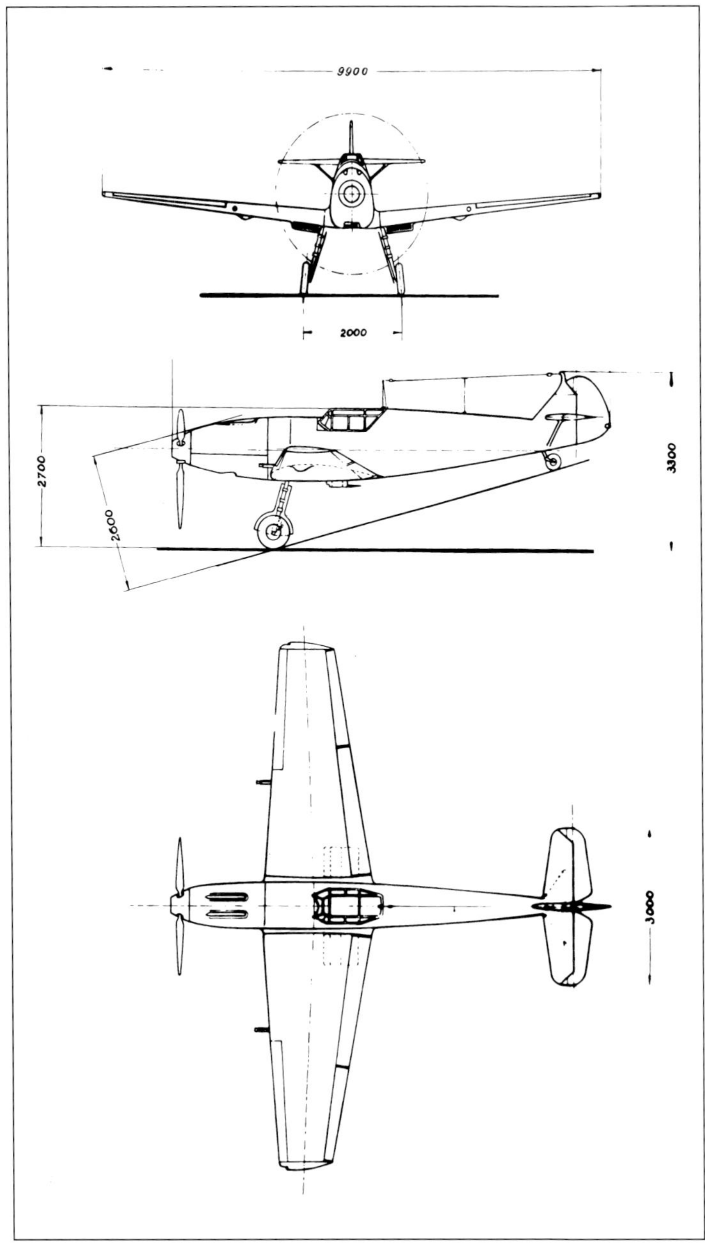


**Three-view drawing of the Bf 109 E-3, the first version to be equipped with two MG 17s and two MG FF wing-mounted cannon.**

The MG FF was positioned in the wing on two mounts (fore and aft) between Ribs 3a and 3b. Each weapon was situated 2,280 mm from the fuselage centerline. The right weapon was positioned in such a way that the ammunition drum was on its left, while that of the left weapon was on its right. The shells were forced out of the sixty-round T 60 FF drums into the cannon breeches by means of springs. The empty shell casings were ejected from the side of the weapon opposite the ammunition feed and were collected in a shell casing area. An access cover provided access to the area and allowed the casings to be removed.

The MG FF cannon were aligned in such a way that their trajectories crossed those of the fuselage MG 17s at 200 meters and the horizontal plane of position of the Revi C/12 C at 400 meters.

The following 35 Bf 109 E-3s were built by the Augsburg factory:

| 1793 | 1927 | 1939 | 1949 |
|---|---|---|---|
| 1794 | 1928 | 1940 | 1950 |
| 1795 | | 1941 | 1951 |
| | 1931 | 1942 | 1952 |
| 1797 | 1932 | 1943 | 1953 |
| 1798 | 1933 | 1944 | 1954 |
| 1799 | 1934 | 1945 | |
| | 1935 | 1946 | |
| 1802 | 1936 | 1947 | |
| 1803 | 1937 | 1948 | |
| 1804 | 1938 | | |

Deliveries from the Regensburg plant began with *Werknummer* 1955.

### The Bf 109 E-3 for Export

The prototype for the Bf 109 E-3a export version was Bf 109 E-3 WerkNr. 1797, D-IRTT (from 1940: CE + BL). Its first flight was on 26 October 1938 with Fritz Wendel at the controls. It was converted into the export version at the beginning of 1939. All systems and equipment not cleared for export were removed. Furthermore, the machine was painted in a special finish, with the wings and fuselage undersides sprayed in a different color. In this configuration it served as a demonstration aircraft for foreign military delegations. It was also used as a test-bed for the Jumo 211 Da, with which variable-pitch propellers by Junkers, Me P 6 and VDM were tested. On 24 June 1941 the machine arrived at the Max Gerner repair facility at Frankfurt/Main to be returned to E-3 production standard, after which further tests were conducted at Augsburg. Incidentally, D-IRTT is incorrectly identified as the Bf 109 V14 in most publications.

The Bf 109 E-3 was the sole export version of the Bf 109 E. Almost all aircraft of this type were built in the Regensburg factory.

At a conference with the *Luftwaffe* Chief-of-Staff it was decided that sales of this type should be strictly limited at first, no more than about 70 aircraft, however, this proved impossible to carry out. An order from the general staff stated that during manufacture and acceptance flights export aircraft were to be specially marked with a red antenna mast to immediately identify them as "specially equipped."

The structure of the E-3a was similar to that of the E-1. It was powered by a DB 601 Aa producing 1,175 H.P. for takeoff. Since this was an engine with direct fuel

One of the few Bf 109 E-3s exported to Japan, seen here in the hands of Kawasaki in Gifu (top of page). Messerschmitt sent Willy Stör to Japan to act as advisor to Kawasaki; here he is seen with Japanese pilots and mechanics (center and bottom).

**Partial List of Bf 109 E-3a Exports:**

| | | | |
|---|---|---|---|
| **Bulgaria** | 10 | 1940 | from the WNF factory |
| | 9 | 1941 | |
| **Japan** | 10 | 1940 | |
| | 2 | 1941 | |
| **Yugoslavia** | 73 | 1939-40 | without wing armament |
| **Romania** | 11 | 1940 | |
| | 39 | 1941 | withdrawn from schools, converted by the AMME-LUTHER-SECK Co. |
| | 15 | 1942 | |
| **Switzerland** | 56 | 1939 | without armament and radios. The 80 |
| | 24 | 1940 | aircraft received by Switzerland were assigned the codes J-311 to J-390. |
| **Spain** | 44 | 1939 | various models, left over from the "Legion Condor" |
| **USSR** | 3 | 1941 | In November 1940 delivery of one aircraft in march, April and May 1941 was agreed upon. Proof of delivery is not available. |

**Aircraft Type Sheet Bf 109 E-1, E-3**
(from L.Dv.556/3)

| **Dimensions:** | Wingspan | 9 900 mm |
|---|---|---|
| Overall length | 8 800 mm | |
| Maximum height, tail down | 2 600 mm | |
| Wheel base | 2 000 mm | |
| Wing area | 16.4 $m^2$ | |
| | | |
| **Weights** | Bf 109 E-1 | Bf 109 E-3 |
| Empty weight | 1 860 kg | 1 865 kg |
| Additional equipment | 169 kg | 188 kg |
| Equipped weight | 2 029 kg | 2 053 kg |
| Fuel (400 l) | 304 kg | 304 kg |
| Oil (29.5 l) | 27 kg | 27 kg |
| Pilot (parachute + special clothing) | 100 kg | 100 kg |
| Ammunition f. MG 17 (2,000 rounds) | 59 kg | 59 kg |
| Ammunition w. MG 17 (1,000 rounds) | 29 kg | |
| Ammunition w. MG FF (120 rounds) | 40 kg | |
| Ballast | 25 kg | 25 kg |
| Useful load | 544 kg | 555 kg |
| Takeoff weight | 2 573 kg | 2 608 kg |
| Max. allowable takeoff weight | 2 610 kg | 2 610 kg |
| Wing loading based on takeoff weight of 2 608 kg | | 159 kg/$m^2$ |

injection, it was initially considered as "secret war equipment" and was excluded from export. Not until the first Bf 109 E-3s and other German warplanes with similar high-performance engines were shot down during the French campaign did the Allies get a look at the German-developed fuel-injection technology. Incidentally, the first Bf 109 E-3 to fall into French hands intact was "White 1" (WerkNr. 1304), which made a forced landing near Woerth (Lower Alsace) on 22 November 1939. After brief trials in France (Orleans-Bricy), in January 1940 it arrived at Boscombe Down and in May went to Farnborough, where it was assigned the British serial number AE 479.

Those aircraft destined for export had to be laid down by the manufacturers outside the Reichs-Program (for the RLM). Only in special cases was it possible for some aircraft to be taken from the production lines for export purposes. Export to "friendly and neutral states" required RLM authorization not just for the aircraft, but also for the engine, instruments, equipment and armament installed in it.

### Bf 109 E-3/B

Fighter or fighter-bomber. Airframe similar to that of the E-3. Power plant: DB 601 A. Was equipped with bomb-dropping system (So-3) as described in the section on the E-1/B (without Point 4).

### Bf 109 E-4

Fighter. This version was equipped with a sturdier canopy framework and an armored head support for protection against enemy fire from the rear. It was otherwise similar to the E-3 with one exception: for the first time MG FF/M cannon were installed in the wings. The "M" meant that the weapon was capable of firing the FFM *Minen-Geschosspatrone*. To this end the following was stenciled on the inside and outside of the ammunition drum access panel: "Attention! MG-FF 'M'!"

The newly-developed *Minen-Geschosspatrone* was a very thin-walled shell with a larger explosive charge, which in combination with a self-destroying fuse produced

| | | |
|---|---|---|
| **Performance:** | Maximum Speed | |
| at | 0 km | 460 kph |
| | 1 km | 480 kph |
| | 2 km | 500 kph |
| | 3 km | 520 kph |
| | 4 km | 540 kph |
| | 5 km | 555 kph |
| | 6 km | 555 kph |
| | 7 km | 550 kph |

| | | |
|---|---|---|
| **Time to Climb:** (rounded off) | to 1 km | 1 min. |
| | 3 km | 3 min |
| | 6 km | 6.3 min |
| | 9 km | 16 min |

**Service Ceiling:** approximately 10,300 m

**Flight Limits:**

| | |
|---|---|
| Maximum allowable horizontal speed | 485 kph |
| Maximum speed with landing flaps down | 250 kph |
| Maximum speed with undercarriage down | 250 kph |
| Maximum diving speed | 750 kph |
| Maximum engine speed in diving flight (30 sec.) | 3,000 rpm |
| Takeoff speed | 140 kph |
| Landing speed on touchdown | 135 kph |
| Takeoff distance to 20 meters | 407 m |
| Landing distance from 20 meters to full stop | 530 m |
| Tightest turning radius at ground level | 125 m |
| Tightest turning radius at altitude of 6 km (with landing flaps at maximum deflection) | 230 m |

**Range***

| RPM | Altitude | Flying Time | Distance |
|---|---|---|---|
| 2,200 | 1 km | 1 hr 5 min | 430 km |
| 1,300 | | 2 hr 20 min | 650 km |
| 2,200 | 3 km | 1 hr | 450 km |
| 1,300 | | 2 hr 5 min | 660 km |
| 2,400 | 5 km | 55 min | 460 km |
| 1,400 | | 1 hr 50 min | 665 km |
| 2,400 | 6 km | 1 hr 10 min | 520 km |
| 1,600 | | 1 hr 40 min | 635 km |

*According to L.Dv.556/3, Appendix 9

**Power Plant** DB 601 A, 12-cylinder in two cylinder blocks each of six cylinders in inverted vee arrangement (60°), injection pump, supercharger for 4 km with independent pressure regulation. Total displacement: 33.9 liters.

**Engine Performance:**
(at ground level)

| | |
|---|---|
| Short-duration, increased (1 min.) | 1,175 H.P. at 2,500 rpm |
| Fuel consumption | 433 l/hr |
| Short duration at ground level (5 min.) | 1,015 H.P. at 2,400 rpm |
| Fuel consumption | 321 l/hr |
| Continuous output, increased (30 min.) | 950 H.P. at 2,300 rpm |
| Fuel consumption | 288 l/hr |
| Continuous output | 860 H.P. at 2,200 rpm |
| Fuel consumption | 260 l/hr |

**Engine Performance:**
(at altitude)

| | |
|---|---|
| Short-duration at 3.7 km (5 min.) | 1,100 H.P. at 2,400 rpm |
| Fuel consumption | 318 l/hr |
| Continuous output, increased, at 4.1 km | 1,050 H.P. at 2,400 rpm |
| Fuel consumption | 297 l/hr |
| Continuous output at 4.5 km | 1,000 H.P. at 2,400 rpm |
| Fuel consumption | 283 l/hr |
| Continuous output, economic, at 3.85 km | 975 H.P. at 2,250 rpm |
| Fuel consumption | 269 l/hr |

**Propeller:**

| | |
|---|---|
| VDM variable-pitch, three blades | |
| Diameter | 3.10 m |

a much greater affect on the airframe of a target aircraft than the standard MG FF ammunition. The shells were loaded into the 60-round drum (T 60-FF) in varying sequences depending on the nature of the mission.

The only differences between the MG FF M and the MG FF, Model B were lighter breech slides and a weaker recoil spring. Consequently, the new weapon was somewhat lighter. The ammunition previously used could not be fired, only the shells listed below:

**Bf 109 E-4/B**

Fighter or fighter-bomber. Airframe similar to E-4, but was equipped with bomb-dropping equipment (see E-1/B).

**Bf 109 E-4/N**

Fighter. Similar to the E-4, but powered by the DB 601 N power plant. The engine was derived from the DB 601 A but differed from it in having a better performance, which was achieved through increased revolutions, higher compression and C3 special fuel (100 octane). The engine entered series production in 1939.

Initially the aircraft equipped with "N" engines were tied to certain bases of operation, since C3 fuel was not yet available everywhere. As well, in some cases a special marking consisting of the letter "N" was stenciled on the engine cowlings of Bf 109s (and Bf 110s) equipped with the DB 601 N (see table above right).

**Bf 109 E-4/BN**

Fighter or fighter-bomber. Identical to E-4/N but with bomb-dropping equipment.

**Bf 109 E-5**

Tactical reconnaissance aircraft. Similar to the E-1 with DB 601 A. For this mission the radio equipment in the fuselage was replaced with a Rb 21/18 camera. The camera operation was electric. The film cassette contained 60 meters of film, sufficient for approximately 300 photographs (18 x 18 cm).

**Bf 109 E-6/N**

Tactical reconnaissance aircraft. Airframe similar to E-1. DB 601 N power plant. Photographic equipment consisted of two HK 12.5/7 x 9 motorized cameras. These were mounted one behind the other between Frames 5 and 6 pointing outward 12 degrees. A sliding hatch prevented fouling of the lenses (by exhaust residue and oil), this being operated from the cockpit by means of a pull cable. Electric controls enabled

**Useable Ammunition**

| | | |
|---|---|---|
| Incendiary high-explosive shell, tracer | FFM | no self-destructing fuse fragmentation, incendiary and explosive effect |
| Incendiary high-explosive shell, tracer | FFM | with self-destructing fuse fragmentation, incendiary and explosive effect |
| Mine shell | FFM | no self-destructing fuse gas explosion (mine) effect |
| Mine shell | FFM | with self-destructing fuse gas explosion (mine) effect |
| Armor-piercing shell | FFM | no self-destructing fuse penetrative effect |
| Armor-piercing shell | FFM | with self-destructing fuse explosive and fragmentation effect after penetrating 5 mm of armor |
| Armor-piercing incendiary shell | FFM | no self-destructing fuse against heavily armored aircraft (phosphorous) |
| High-explosive shell, practice | FFM | no self-destructing fuse weapon function checks |
| Armor-piercing shell, practice | FFM | no self-destructing fuse acceptance firing |
| High-explosive shell, tracer, practice | FFM | no self-destructing fuse practice firing |
| High-explosive shell, tracer, practice | FFM | with self-destructing fuse practice firing in confined firing areas (L.Dv.4000, Part 10) |

**DB 601 N Engine Performance:**

| | | |
|---|---|---|
| Maximum output at | 2,600 rpm at ground level | 1,175 H.P. (1 min.) |
| Maximum output at | 2,600 rpm at 4.9 km | 1,175 H.P. (1 min., max. boost alt.) |
| Short-duration output at | 2,400 rpm at ground level | 1,020 H.P. (20 min.) |
| Short-duration output at | 2,400 rpm at 4.8 km | 1,050 H.P. (30 min.) |
| Continuous output at | 2,300 rpm at ground level | 910 H.P. |
| Continuous output at | 2,300 rpm at 5.1 km | 950 H.P. |

the pilot to select strip or single frame photography. One film made it possible to expose sixty negatives (7 x 9 cm).

### Bf 109 E-7/B

Extended range fighter or fighter-bomber. After the limited range of the "E" had proved a serious disadvantage in operations over Great Britain, this version introduced a special carrier for a jettisonable 300-liter fuel tank, which was exchangeable for an ETC 500.

In order to avoid difficulties in the allocation of new aircraft, it was decreed that at least the first fifty aircraft delivered with the external tank were all to go to the same front-line unit.

### Bf 109 E-7/N

Extended range fighter. In this variant the oxygen system was increased from two to three bottles, plus an additional bottle (2 liters). At the same time the Dräger oxygen apparatus was replaced by the standard

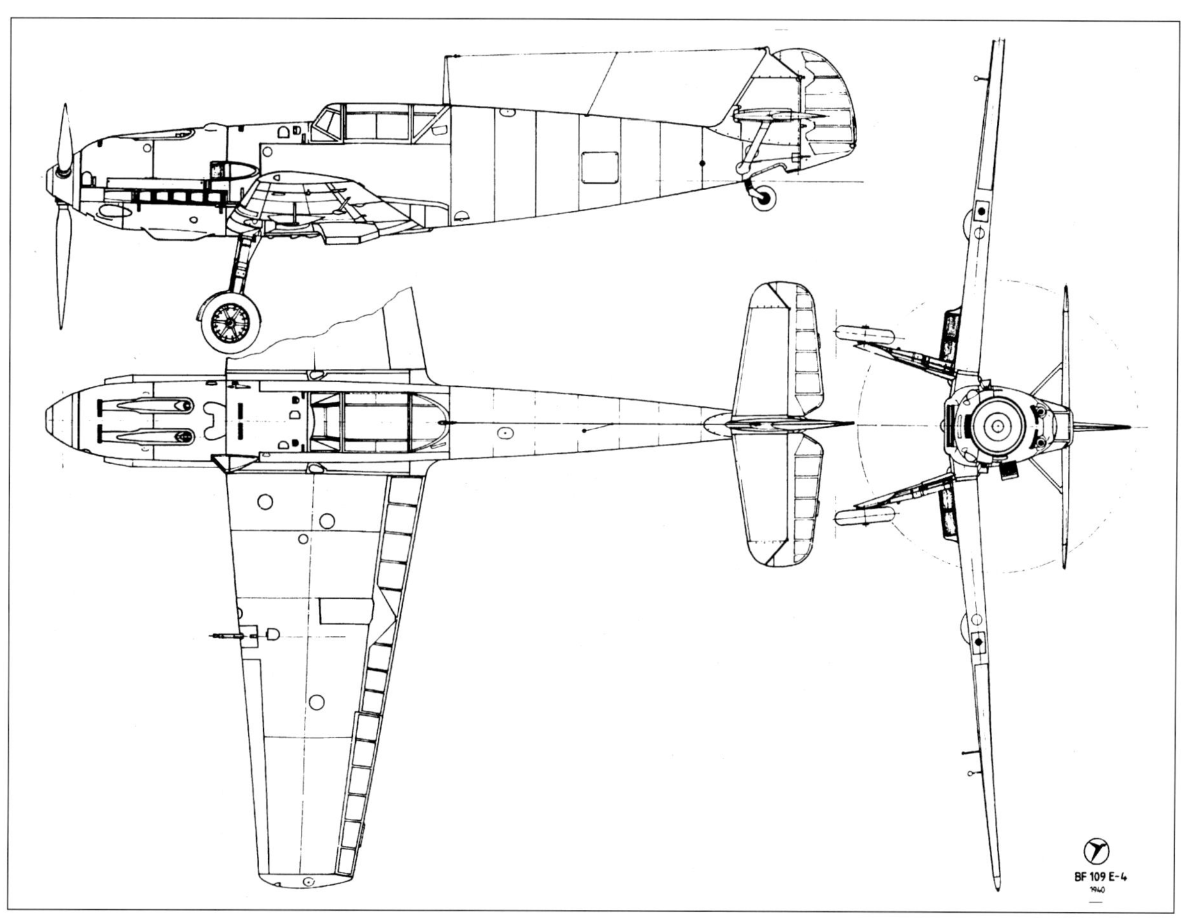


Three-view drawing of the Bf 109 E-4, which was built as a fighter and a fighter-bomber. (Sengfelder)

Right: One of the first Bf 109 E-3s with two fuselage-mounted MG 17s and two MG FF cannon in the wings. Below: Bf 109 E-3 of the Swiss *Flugwaffe*.

Opposite page: Cockpit of a Bf 109 E-3 fighter-bomber; this machine still has the Elektron rudder pedals, which in later versions were replaced by standard pedals.

phragm breathing system. While the three oxygen bottles supplied the breathing system, the additional bottle was connected to the oxygen tube by way of a pressure reducer and acted as a blower system (high-pressure breathing apparatus). After opening the stop valve, the pilot received pure oxygen. For safety reasons, pure oxygen without atmospheric air mixed in had to be used at altitudes between 8,000 and 12,000 meters. Use was limited to twenty minutes. (D.(Luft)T.291/3, March 1941).

## Bf 109 E-7/Z

Extended range fighter or fighter-bomber. Power plant: DB 601 N with GM 1 system. The suffix "Z" indicated that the machine was equipped with a GM 1 system ("Mona"). The system's purpose was to increase engine output above its maximum boost altitude. The increase in performance shown by the DB 601 N after switching on the GM 1 system (100 g/sec) was 250 to 280 H.P. at an altitude of 8,000 meters, which resulted in a speed increase of about 100 kilometers per hour. The GM 1 system injected oxygen into the engine. The oxygen was carried by nitrous-oxide ($N_2O$, or "laughing gas"), which could be injected at various rates (60, 100 and 150 g/sec). The GM-1 material was a liquefied gas (by pressure or cold) of about -90° C; compressed air from three insulated air bottles under 4.5 kg/m$^2$ of pressure forced the gas through insulated lines to the atomizer nozzle in the supercharger throat. There the GM 1 material entered the intake air stream, was atomized and vaporized, and thus fed into the cylinders with the supercharger air. The system could not be used below an altitude of 6,500 meters. Use of the system raised fuel consumption. The

D2
C11
P15
M7
P14
Nullschuß
E10
Ein
Aus
mv
mv
Sturz
OV
OV
ata
U/min
Liter
°C
°C

**DB 601 on the hoist during installation on the production line.**

GM 1 system could be installed on the production line or retrofitted to existing aircraft. Dortmund was named as the retrofit center. The first three *Staffeln* equipped with GM 1 were from the *Jagdgeschwader* commanded by Obstlt. Mölders, Obstlt. Galland and *Major* Helmut Wick.

### Bf 109 E-8

Extended range fighter. Structure similar to the E-7, but based on the E-1 (DB 601 A).

**Bf 109 E-7 Conversion Designations** (U = conversion)

| | | |
|---|---|---|
| Bf 109 E-7/U1 | Armored radiator | 208 examples[1] |
| Bf 109 E-7/U2 | Close-support armor and SG-tank[2] | 265 examples[2] |
| Bf 109 E-7/U3 | Fitted with two staggered Rb 12.5/7 x 9 (as E-6/N) and FuG 17 (VHF radio) | |

[1] Target number from conversion program of 30 November 1942
[2] SG = protected bag-type fuel tank, contents 400 liters.

### Bf 109 E-9

Extended range reconnaissance aircraft. This variant was equipped with Rb 50/30 cameras for use at higher altitudes. The film cassette contained 120 meters of film, with which approximately 380 photographs (30 x 30 cm) could be taken.

### Operations in the Tropics

A number of modifications had to be made to the Bf 109 E prior to its use in the tropical conditions of North Africa. All of these were undertaken as retrofits, since there was no production E/trop. The most prominent addition was the sand filter in front of the supercharger air intake. Since tropical versions of the Bf 109 did not begin leaving the production lines until 1 June 1941 (F series), several E-variants (from the E-4) had to be modified for the new role of tropical fighter, for which sufficient conversion parts were available. I/JG 27 was the first unit to be equipped with the Bf 109 E/trop, receiving fifty aircraft for its four *Staffeln* in April 1941. The production of E/trop machines by the repair industry was increased to thirty aircraft per month to ensure that the four *Staffeln* were kept supplied until 1 October 1941.

It is a little-known fact that the standard tropical equipment included a Kar 98 K carbine, which was attached to the left interior wall of the aft fuselage. There was also a sun umbrella, which was supposed to prevent cockpit overheating in aircraft parked in takeoff position (scramble flights). According to D.(Luft)T.2109, Part 9F, the sidewalls of the tires were supposed to be painted in a special white reflective finish.

The Bf 109 E production line.

### Bf 109 E Production Totals

It is very difficult, if not to say impossible, to determine the precise number of Bf 109 E series aircraft built. There are several reasons for this:

- The absence of production and acceptance records from Messerschmitt and those companies that built the aircraft under license.
- In spite of the many surviving delivery plans and C-Amts programs, they are not reliable enough for the determination of a definite final figure, as a comparison of individual delivery programs reveals differences which do not allow a precise accounting. As well, continuous changes to the delivery plans and programs resulted in postponements of individual variants, which affected both the number of aircraft to be delivered and the variant. In mid-1940 the units rejected the E-1 because it was inferior in action. Consequently, beginning in August 1940, the remaining E-1s in service were converted into E-4s or E-7s (DB 601 N, heavier armament), while the last 175 E-1s (according to the delivery plan) were completed as E-7/Ns.

In order to provide a certain idea of the scope of Bf 109 E production, there follows a delivery plan (the months November and December 1940 are missing) as well as production figures for 1941.

BY

Series production of the Bf 109 E in Messerschmitt's Regensburg plant: fuselage assembly (above), joining of wings and fuselage (above left), and undercarriage assembly (left).

Rauchen

Following final assembly (above) the aircraft were rolled out of the hall of the Regensburg factory (left and above left).

Visierdatenblatt für die beiden gesteuerten MG17

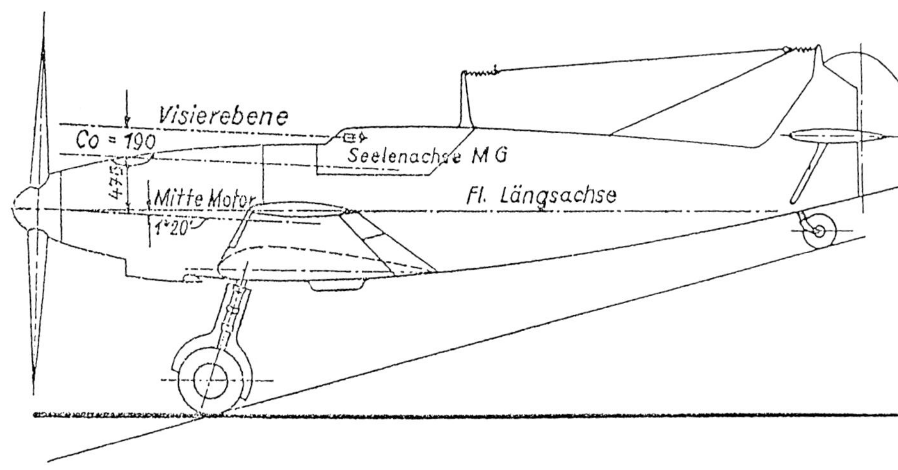

Aligning mechanism for the MG FF prior to adjustment fire (above left and above); sighting data sheet for the two synchronized MG 17s (left).

Optical device (top of page) for aligning the MG FF (above). Bf 109 E-3 demonstration aircraft, which was dubbed "Cockatoo" on account of its colorful paint scheme (far left).

**Delivery Plan Amendment Sheet B No.18/3 No.1285/40 Secret Command Matter. C-Amts Program of 1/11/1940**

| | | Fighters to 31/10/40 | | |
|---|---|---|---|---|
| Bf 109 E-1 | DB 601 A | Mtt.A | 14 | 14 |
| | | FW | 90 | 90 |
| | | Ago | 80 | 80 |
| | | Fi | 447 | 447 |
| | | Ar.W | 442 | 442 |
| E-1/B | DB 601 A | Fi | 61 | 61 |
| | | Ar.W | 49 | 49 |
| | | | 1183 | 1183 |
| Bf 109 E-3 | DB 601 A | Mtt.A | 35 | 35 |
| | | Mtt.R | 75 | 75 |
| | | Erla | 812 | 812 |
| | | WNF | 249 | 249 |
| E-3a | DB 601 A | Mtt.R | 75 | 75 |
| | | | 1246 | 1246 |
| Bf 109 E-4 | DB 601 A | Mtt.R | 32 | 32 |
| | | Erla | 99 | 99 |
| | | WNF | 119 | 119 |
| E-4/B | DB 601 A | Mtt.R | 35 | 35 |
| | | Erla | 96 | 96 |
| | | WNF | 80 | 80 |
| E-4/BN | DB 601 N | Mtt.R | 15 | 15 |
| | | WNF | 20 | 20 |
| | | | 496 | 496 |
| Bf 109 E-5 | DB 601 A | Ar.W | 29 | 29 |
| Bf 109 E-6/N | DB 601 N | Ar.W | 9 | 9 |
| Bf 109 E-7/N | DB 601 N | BFW.R | 63 | 63 |
| | | Erla | 106 | 106 |
| | | Fi | 135 | 135 |
| | | Ar.W | 46 | 46 |
| | | WNF | 102 | 102 |
| | | | 452 | 452 |
| Bf 109 E-8 | DB 601 A | Fi | 22 | 22 |
| | | Ar.W | 38 | 38 |
| | | | 60 | 60 |
| Bf 109 E-9 | DB 601 N | Is contained in the figures with DB 601 N engines, the type was a conversion into an extended-range reconnaissance aircraft (only a few completed). | | |

**Deliveries of Bf 109s in 1941**

| Month | Firm | in month Delivered | Authorized | Actual | Type |
|---|---|---|---|---|---|
| January | Ar.W | 1 | 9 | 9 | E-6/N |
| | WNF | 0 | 82 | 72 | E-7/N |
| | AR.W | 3 | 33 | 36 | E-7/N |
| | Fi | 19 | 117 | 69 | E-7/N |
| | Erla | 9 | 105 | 105 | E-7/N |
| | Mtt.R | 1 | 63 | 62 | E-7/N |
| February | WNF | 0 | 92 | 72 | E-7/N changed to 92 |
| | Ar.W | 8 | 46 | 44 | E-7/N changed to 46 |
| | Fi | 20 | 135 | 89 | E-7/N |
| | Mtt.R | 0 | 63 | 62 | E-7/N |
| March | WNF | 5 | 87 | 77 | E-7/N changed to 87 |
| | Ar.W | 2 | 46 | 46 | E-7/N |
| | Fi | 30 | 135 | 119 | E-7/N |
| | Mtt.R | 1 | 46 | 46 | E-7/N |
| | 16 | 16 | 62 | E-7/NZ | |
| April | WNF | 4 E-4/NZ | 87 | 81 | |
| | Fi | 16 E-7/NZ | 135 | 135 | E-7/N |
| May | WNF | 0 | 87 | 81 | E-7/N + E-7/NZ |
| June | WNF | 6 E-7/NZ | 87 | 87 | E-7/N + E-7/NZ |

Production of the Bf 109 E by all firms ran down between January and March 1941. The last machines were accepted by the *Luftwaffe* in June 1941. A reasonable estimate of total production is approximately 4,000 machines.

**Production of the Bf 109 was distributed as follows:**

Messerschmitt AG, Augsburg: E-0, E-1, E-3
Messerschmitt GmbH, Regensburg: E-1, E-3, E-3a, E-4, E-4/BN, E-7
AGO AG: E-1
Arado GmbH, Warnemünde factory: E-1/B, E-5, E-6/N, E-7, E-7/N, E-8, E-9
Erla GmbH: E-3, E-4, E-4/B, E-7, E-7/N
Fieseler GmbH: E-1, E-1/B, E-7, E-7/N, E-7/NZ, E-8, T-2
Wiener-Neustädter Flugzeugwerke GmbH: E-3, E-4, E-4/B, E-4/BN, E-7, E-7/N, E-7/NZ

# Bf 109 Carrier Aircraft

Development of an aircraft carrier also included special aircraft suited to carrier operations. Since its story has been examined in detail in other publications, no space will be devoted here to the first German aircraft carrier "A," the *Graf Zeppelin*, whose keel was laid at the "Deutsche Werke Kiel" (DWK) shipyards on 28 December 1936 and who was launched on 8 December 1938.

The aircraft carrier was mentioned as early as 11 May 1934 during a conference in the *C-Amt* on "Tactical Requirements and Urgency of Aircraft Development Tasks." At that time, however, there was still no concrete idea of the aircraft type to be developed for it. The requirement led to a "carrier fighter-dive bomber aircraft."

The first "Employment Guidelines for Carrier Aircraft" (Chef der Mlt.A.I. L 4753/34 g.Kdos) appeared on 5 December 1934. This was followed on 7 February 1936 by new "Employment Guidelines and Tactical Requirements for Carrier Aircraft (memo LC II 205/36 g.Kdos of 11 January 1936). In it the guidelines of 1934 were modified based on the "Japan Committee 1935." The requirement (excerpt) now stated: "to build the land-based fighter as a carrier fighter, likewise the land-based dive bomber as a carrier dive bomber." The carrier fighter-dive bomber aircraft remained the ultimate requirement, however, and was to continue as a study. The carrier fighter was to have a maximum speed of 400 kph at an altitude of 6 kilometers and a landing speed of less than 100 kph. Furthermore, the fighter was required to be capable of 1 1/2 hours at full power at an altitude of 6 kilometers. Specified time to climb to six kilometers was seven minutes. Service ceiling was ten kilometers. As well, the aircraft was to be equipped with lift-destroying devices (spoilers), arrestor hook, wheel brakes and catapult fittings. In February 1936 working information was issued to the Development Group LC II/1d for a carrier-based land aircraft (multi-purpose), which resulted in the issuing of development contracts to Arado and Fieseler in March 1936. The contracts contained the directions for "Multi-Purpose Carrier Air-

**The Bf 109 D for arrested landing trials was test flown by Fritz Wendel on 18 March 1938 and on 25 July was ferried to Travemünde. Clearly visible are the catapult fittings and the arrestor hook on the fuselage underside as well as the arrestor cable fender in front of the mainwheels.**

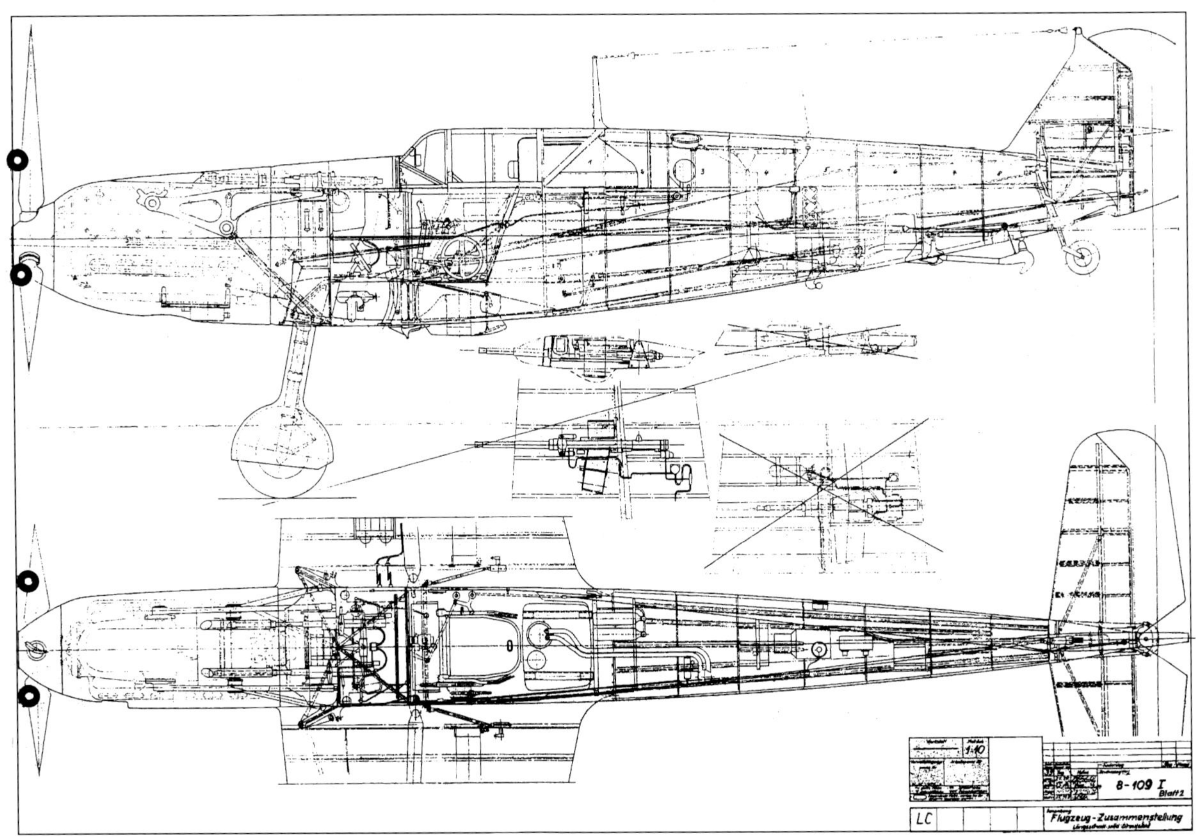


**General arrangement drawing of the Bf 109 T with catapult fittings and arrestor hook.**

craft 36" (Ar 195, Fi 167). The carrier fighter aircraft was to be developed from the pursuit fighter which was to be procured for the *Luftwaffe*. This led to the Bf 109 T. The same applied to the carrier Stuka, which was a variant of the Ju 87. The *E-Stelle Travemünde* used a number of experimental aircraft for catapult and arrested landing trials from 1938 until the cessation of carrier development at the beginning of 1943 (see table below).

The *E-Stelle* had to carry out numerous tests for the "carrier aircraft" mission area: in addition to carrying out the usual flight tests (performance, handling, range), it was required to evaluate all specialized carrier equipment and conduct extensive catapult takeoff and arrestor landing trials. For arrested landings a combined electro-mechanical carrier arresting system by the Atlas Werke company was installed, however, in 1939 this was replaced by a DEMAG system. Landing required the pilot to make a precise three-point landing in which the arrestor hook—hanging almost vertically from the rear fuselage—engaged the arrestor cable, allowing the aircraft to be stopped by the arrestor gear. A landing surface equal in size to what would be available on the carrier (approx. 22 meters

**Experimental Aircraft**

| | |
|---|---|
| Bf 109 E and T | Carrier-based fighter aircraft |
| Ju 87 C | Carrier-based Stuka and torpedo bomber |
| Fi 167 | Carrier-based reconnaissance aircraft (multi-purpose) |
| Ar 195 | Carrier-based reconnaissance aircraft (multi-purpose) |
| Ar 197 | Carrier-based fighter trainer |
| Ar 96 B | Carrier training aircraft |
| He 50 | Arrested landing tests |
| Avia 534 | Arrested landing tests (proved structurally unsuited) |

Note: The Ju 87 C, Fi 167 and Ar 195 were equipped with folding wings on account of the size of the carrier's hangar elevators.

long) was marked out on the landing field ("carrier mock-up"). The braking distance was between 20 and 30 meters. Approximately 1,800 arrested landings were carried out with the Travemünde arrestor gear with no fatal accidents. Catapult takeoffs were made from the *E-Stelle* dock catapult. The aircraft was first lifted by a crane onto the catapult slides, to which the four catapult fittings were attached, then it was moved onto the catapult.

**Tests at the *E-Stelle Travemünde* in preparation for operations from the aircraft carrier Graf Zeppelin: Bf 109 V17 with arrestor hook and catapult fittings, arrestor cable deflector in front of the tailwheel (above).**
**Bf 109 B after catapult launch from the floating dock (above). The aircraft has no arrestor hook but does have catapult fittings. Note the landing flaps in "takeoff position". The photograph was taken in 1940. The code TK + H stood for the *E-Stelle Travemünde*, M was the aircraft letter.**

**Bf 109 Carrier Trials Aircraft Used by the *E-Stelle Travemünde***
**Bf 109 V17, Werk.Nr. 1776, D-IYMS** (TK + HK)

First flight on 24 February 1938 with Fritz Wendel at the controls. The 1 1/2 hour flight was completed on 18 March. Aircraft was powered by a Jumo 210 D, armament consisted of two fuselage-mounted and two wing-mounted MG 17s operated electrically.

The V17 was the first machine converted for use as a carrier aircraft in Messerschmitt's prototype shop. Various reinforcements were made to the aft fuselage, and the mainwheels and tailwheel were fitted with cable deflectors. Also fitted were catapult fittings and an arrestor hook. The only feature of the carrier version that was lacking was the extended wing. In May 1938 the aircraft was ferried to Travemünde, and on 25 July 1938 it was damaged in a takeoff crash. Consequently, there were delays in the Bf 109 carrier program. After it was repaired, the machine was assigned the code TK + HK at the *E-Stelle Travemünde*.

**Bf 109 V17a, WerkNr. 301, D-IKAC** (TK + HM)

Built by Erla. Acceptance by BAL officials in Augsburg took place on 25 July 1938. The aircraft initially served as a test-bed for the DB 601 A. Then it was equipped for carrier trials and fitted with a Jumo 210 D. The usual tests were carried out at Travemünde, in the course of which the cable deflectors proved unnecessary.

At the beginning of 1939 Messerschmitt-Augsburg carried out experiments whose objective was to find the most suitable wing for the carrier aircraft. The following modifications were made for this purpose:

- The Bf 109 V15 received a new wing which was lengthened by approximately 1.2 meters, with correspondingly longer ailerons and leading edge slats. Wing area was increased by approximately 1.5 m$^2$.
- A Bf 109 B was fitted with a wooden wing with 3° aerodynamic dihedral and fixed slats. The aircraft's weight was increased to that of the Bf 109 E (2,460 kg).
- A Bf 109 F was fitted with ailerons projecting beyond the trailing edge of the wing. A Bf 109 E from current production was acquired for comparative trials.

Results:

- The lengthened wing proved significantly superior to the standard wing in the stall and in aileron effectiveness at low speeds. This wing was selected for the production Bf 109 carrier aircraft.
- While the new wing was superior to the E wing, it stalled more forcefully and considerably sooner.
- The F wing stalled forcefully and was unusable for carrier use.

**Bf 109 V15, WerkNr. 1773, D-IPHR** (CE + HL)

The converted V15 (extended wing, arrestor hook, spoiler, etc) also took part in the tests. It was still flying in 1942.

**Bf 109 E-0, WerkNr. 1781, D-IECY** (WL-IECY, TK + HL)

The aircraft was produced in Augsburg and was equipped with a DB 600 G power plant. Dr. Würster carried out the first flight on 4 August 1938. It was converted for arrested landings, in the course of which the radio and antenna mast were removed, as were the catapult fittings. The arrestor hook was an experimental model.

The machine underwent extensive testing in Augsburg (including cockpit heating). On 29 September 1939 Fritz Wendel ferried it to the *E-Stelle Rechlin*, and on 11 October returned with it to Augsburg. A shortened (2.9 meters) three-blade VDM variable-pitch propeller was subsequently installed. In may 1939 it went to Travemünde for further tests (including coolant heating).

**Bf 109 E-0, WerkNr. 1783, GH + NT)**

Built at Augsburg. The exact date of its first flight is not known, but it can be assumed to have been September 1938. Installed power plant was a DB 601 A. Equipped with an arrestor hook, it was used for arrested landing trials and was demonstrated to Udet during a visit by him to Travemünde.

**Bf 109 E-3, WerkNr. 1946, D-IPGV** (GH + NU)

Also built at Augsburg. Fritz Wendel conducted the first flight on 16 December 1938 and also ferried the machine to the *E-Stelle Rechlin* on 27 July 1939. From there the machine went to Travemünde for arrested landing trials. In July 1940 it was flown back to Augsburg for installation of the Me P 6 reversible-pitch propeller and trials with same. These were carried out by Karl Baur in August 1940. Propeller trials continued at Travemünde from 16 October 1940.

**Bf 109 WerkNr. 4950, VG + CX**

One interesting experiment was "braked landings" involving the Messerschmitt-developed Me P 6 propeller. The propeller was capable of reversing thrust, making it possible to taxi the aircraft backwards. In May 1940 the *E-Stelle Rechlin* carried out a series of "braked landings" in which the roll-out distance of the Bf 109 was reduced to a maximum of 100 meters. They were convinced that the effect could be further improved and the landing run further shortened through appropriate measures, for example, increasing the negative pitch angle blade of the propeller blades. This naturally made it an attractive feature for carrier landings.

**Bf 109 WerkNr. 6153, CK + NC**
**Prototype for the T-Series**

The date of the aircraft's first flight cannot be determined since it was a license-built machine. After being ferried to Augsburg for a stopping and dynamic pressure calibration flight, on 15 May it completed its 1 1/2 hour flight in the hands of Dr. Würster.

**Bf 109 E (WL + IECY) at *E-Stelle Travemünde* for arrested landing trials in September 1939. It has an experimental arrestor hook, no catapult fittings and no antenna mast. Code in April 1940: TK + HL.**

Then began the conversion into the prototype, which was equipped as follows:

- Increased wingspan from 9.90 m to 11.08 m, resulting wing area 17.5 m$^2$.
- Span with wings folded 4.59 m.
- Arrestor hook with articulated mounting and release cable for the hook.
- Catapult fittings, 2 forward and 2 aft of the attachments for the four-point takeoff trolley, one spoiler on the upper surface of each wing.
- Armored pilot seat.
- Night lighting.
- PATIN remote-reading compass.

- Seat parachute with one-man dinghy.
- Arm rests and thicker headrest on the back armor.

After being test flown several times the machine was sent to Travemünde for further testing, where on 6 August 1940 GFW test pilot Anton Riedinger carried out arrested landing trials.

### Bf 109 T (carrier aircraft)

Based on test results (catapult launches and arrested landings) it was now obvious that the modified Bf 109 E-7/N was the best candidate for the carrier fighter. The only remaining doubts concerned the aircraft's landing characteristics on the carrier. The following is from a letter from the RdL. and ObdL to the OKM on 13 September 1938:

"The Bf 109 is acknowledged to be the best single-seat fighter type. In the opinion of the L.In.8 it does not appear justifiable to employ a type with a poorer performance and larger dimensions, such as the Ar 197 for example, for this purpose (RLM is to report on the Bf 109's carrier landing characteristics."

Immediately a secret command matter was received from the 1.Abt.Skl.I dated 20 September 1938 which was worded as follows (excerpt):

"To answer the question of using the Bf 109 single-seat fighter on board the carrier, fundamentally there are no reservations about landing this aircraft type into a 30 kph wind (equivalent to a carrier speed of 18 knots in still wind conditions).

However, before deciding to use the Bf 109 as a carrier-based single-seater it is asked that the Bf 109's carrier landing characteristics be investigated, since it is known that this aircraft type tends to swing on landing, which on board a carrier would entail a serious threat of crashing after contact with the carrier deck and failure to engage the arrestor cable immediately."

It was decided to wait for the results of arrested landing experiments using the stationary test arrestor system. If the trials at Travemünde revealed this aircraft type to be unsuitable, a request for the design of a new, fast carrier fighter aircraft would have to be considered, since in the opinion of the 1.Skl. the interim operational type, the Ar 197, was unsuitable on account of its relatively modest performance, especially speed and rate of climb.

On 2 December Fl.-Stabsing. Hubrich wrote to the L.In.8: "The OKM has been wrongly informed about the Bf 109's landing characteristics. In no way does the Bf 109 tend to swing on landing unless there is serious pilot error, for which the pilot, not the aircraft, is to blame. Deck landing the Bf 109 will present difficulties for other reasons, mainly because a shallow approach to the deck at a high angle of attack is just as impossible as a steep descent in three-point attitude (as with the Ar 68, for example). The LC therefore expects that specialized training will be required for engaging the arrestor cable in deck landings by the Bf 109."

The Bf 109 (Tr) appears for the first time in Delivery Program No. 11 of 11 April 1939 (Nr.205/39g.Kdos.). Production of sixty examples from the E-3 series at Augsburg was anticipated. The target date for conversion to the "T" was to be determined later. This changed, however, for on 10 September 1939 the OKM declared that the target date for the carrier's entry into service was to remain 1 June 1940. This made the accelerated production of the Bf 109 T urgently necessary, likewise an acceleration of work on the Fi 167 and Ju 87 C.

On 13 October the commander-in-chief of the *Luftwaffe* received the following letter from the *Chef Skl.*:

"No possible use seen for the aircraft carrier *Graf Zeppelin* at this time in the present war. Further construction to be delayed in favor of more pressing mobilization tasks by the navy (new-build ship plan, U-boat program and necessary repairs to the naval forces). Service entry of the aircraft carrier is not to be expected before 1 October 1940, construction will not be accelerated." The *Luftwaffe* General Staff subsequently withdrew the contracts for all of the carrier's aircraft, and on 18 December 1940 canceled further production of the Bf 109 T-1 (with the exception of seven examples).

Meanwhile, planning by the GUC concerning the Bf 109 T went on in spite of the uncertain carrier situation. Thus, for example, Aircraft Procurement Program No. 16 of 25 October 1939 (Nr.811/39g.Kdos.) still called for 170 examples of the Bf 109 T (probably inclusive of the aircraft for Carrier "B," work on which was suspended even earlier than Carrier "A"). The definitive figure of 70 machines appeared in *C-Amts* Program No. 18 of 1 July 1940.

With the two Messerschmitt factories in Augsburg and Regensburg working at full capacity, program No. 17 initially called for Erla to build 100 aircraft and Fieseler 70. From various inquiries made of Messerschmitt-Augsburg by both companies, it is obvious that neither Fieseler nor Erla knew exactly how many aircraft were to be built and whether both firms were supposed to build the Bf 109 T. According to Program No. 17 the total was divided between Erla and Fieseler. After a request for clarification from Messerschmitt to the LC II/Ia, a decision was made that Fieseler alone should build the T-series. Whether 100 or 70 aircraft would be built was yet to be determined. Finally, the figure of 70 aircraft was decided upon (WerkNr. 7728 to 7797). On 16 February 1940 the Commander-in-Chief of the Navy decided to immediately halt further work on the aircraft carrier. He justified this decision as follows:

"Since the heavy losses suffered by the *Kriegsmarine* in the Norwegian operation make it necessary to concentrate all available resources to build smaller units (destroyers, mine-layers and submarines), the naval command regrettably sees itself forced to halt further work on the aircraft carrier *Graf Zeppelin*.

Consequently, the requests made of the *Luftwaffe* to build the anticipated carrier aircraft types are rendered unnecessary. The necessity of building the Bf 109 T aircraft for the operational fighter base under construction on Helgoland is noted."

This decision made construction of the pure carrier version, the Bf 109 T-1, superfluous. Only seven aircraft, *Werknummern* 7728 to 7734 (the first aircraft of the T-series), were to be built as carrier aircraft; the remaining 63 were designated T-2 and were delivered as conventional fighter aircraft.

Installed in the first aircraft (WerkNr. 7728, RB + OA) was a new electrical belt feed system for the two wing-mounted MG FF cannon, which offered increased ammunition capacity over the drum system. The aircraft was flown to the *E-Stelle Tarnewitz*, however, the trials were abandoned in July 1941 after the system failed to achieve a satisfactory level of reliability. Installation of the system in the Bf 109 was rejected, and the sixty-round drums continued in use.

The type also encountered wing flutter, consequently in March 1941 the machines built to date could not be delivered. The search for the cause took place in Augsburg. It was assumed that the flutter was caused by weakly mounted wingtips. The wingtips of WerkNr. 733 (RB+OF) were thickened. The flutter reappeared while the aircraft was in a dive at 750 kph, causing the wing to fail. The machine crashed while being flown by Fritz Wendel on 3 April 1941, but fortunately the pilot was able to escape by parachute.

On 30 May 1941, flying WerkNr. 7743 (RB+OP), Karl Baur demonstrated that the BAL's requirement for a speed of 750 kph at an altitude of 3 000 meters was achievable. In his dive he reached a speed of 760 kph at an altitude of 3,200 meters (outside air temperature -4°).

The Bf 109 T series was delivered as follows:

| 1941 | Accepted by BAL | Total Authorized | Actual |
|---|---|---|---|
| Jan. | — | — | — |
| Feb. | 1 | 10 | 1 |
| March | 0 | 10 | 1 |
| Apr. | 20 | 40 | 21 |
| May | 20 | 60 | 41 |
| June | 29 | 70 | 70 |

End of Bf 109 T production

Testing at Travemünde continued while the aircraft was in production. Performance measuring flights were carried out in April-May 1941, and the following values were achieved ("indicated" = not yet checked):

| | |
|---|---|
| Service ceiling | 11 000 m indicated |
| Time to climb to 22 min. | 11 000 m indicated |
| Time to climb to 16 min. | 10 000 m indicated |
| Speed at 310 kph indicated | 10 000 m indicated |

In May 1941 engine output of the DB 601 N had to be restricted, to 100% = 1.25 atm boost, 2,400 rpm at maximum boost altitude, 2,600 rpm above maximum boost altitude.

Results:

| | | | |
|---|---|---|---|
| 0 m | = | 475 kph | |
| 2 000 m | = | 1.9 min | 515 kph |
| 4 000 m | = | 3.8 min | 555 kph |
| 6 000 m | = | 6.4 min | 575 kph |
| 8 000 m | = | 9.7 min | 560 kph |
| 10 000 m | = | 15.2 min | 520 kph |
| Service ceiling | = | 11 250 m in 24 minutes | |

At the conference held by the *GL 1/ Genst. 6. Abt.* on 11, 15 and 23 May 1941 it was requested that six aircraft (WerkNr. 7728-7732 and 7734, WerkNr. 7733 having been destroyed on 3 April 1941) be equipped to T-1 carrier standard and then stored engineless at the LZA Travemünde until required. The power plants were to go into the supply reserve. The remaining T-2 aircraft went via the LZA Neumünster to Norway, where they saw action (*Jagdgruppe Drontheim*, JG 77, equipped with FuG 16 and FuG 25a).

The situation of the Bf 109 T aircraft underwent a change, however, when during a Navy High Command conference with Hitler on 25 July 1941 the latter ordered the completion of the aircraft carrier *Graf Zeppelin* (earliest completion date 1 October 1942).

The letter by the Ob.d.L. Genst.6.Abt. Nr.7432/41 g.Kdos. (III A) of 23 December 1941 read:

"On order of the *Reichsmarschall* all aircraft suitable for operation from the aircraft carrier are to be readied. The Bf 109 T-2s in front-line service are to be withdrawn and converted into T-1s. Being made available are:

- 5 Bf 109 T-1s, in storage LZA Erding
- 16 Bf 109 T-2s, in Fieseler repair facility, Kassel (of *Luftflotte* 5)
- 30 Bf 109 T-2s, to be released to Fieseler by *Luftflotte 5*."

Of the above-named T-2 aircraft, by 31 December 1942 the Gerhard Fieseler Werke had repaired 46 and converted them to T-1 standard. Acceptance flights took place at Kassel throughout all of 1942. The aircraft were initially assigned to the Gen.d.Lw.b.Ob.d.M. The first went to the LZA Erding, and from there the aircraft were ferried to the *Seeflieger* base at Pillau, and not to Wesermünde air base.

On 19 August 1942 three Bf 109 T-1s sustained serious damage while landing at Pillau; *Luftgau I* was forced to have the runway inspected and temporarily ban the airfield for use by Bf 109s. The subsequent inspection was positive, however, and it was decreed that Bf 109 Ts being ferried to Pillau should first land at Seerappen where a specially selected pilot would take over the aircraft and land it at Pillau. The entire ferry process was handled by *Überführungs-Kommando 12/13*. The airfield at Pillau was earmarked for training pilots to operate from carriers. However, the airfield had been declared unusable by the *General der Jagdflieger*, and the training was now to be shifted to Seerappen, which after installation of an arrestor system would meet requirements. On 29 March 1943 forty-six Bf 109 T-1s were parked on the airfield at Pillau.

In the meantime, however, the OKM had decided to abandon the carrier program for good (on 2 February 1943). As a result, all work to expand the airfields at Pillau and Seerappen was suspended.

The specialized carrier equipment was removed from the Bf 109 T-1s, and effective immediately they were placed at the disposal of the *General der Jagdflieger*, who was to regulate the use of these aircraft from the island of Helgoland (*Düne*) with the *Luftwaffe* Commander Center. The last twelve back-modified Bf 109 Ts were ready to be accepted at Pillau from 28 April 1943. Some went to *Jagdstaffel Helgoland*, while others went to NJG 101 for training purposes and to other units.

**Inflatable dinghy stowed behind the headrest of the Bf 109 T.**

At the behest of the *Technische Amt* the Messerschmitt company had been working on a new carrier aircraft since the beginning of 1942. Work on the "carrier-based single-seat fighter with good high-altitude performance" proceeded side-by-side with work on the Messerschmitt high-altitude fighter (P 1091). Initially designated the Bf 109 ST (ST = "Special Carrier Single-Seater"), or Me 409 (with Jumo 213), the project later received the designation Me 155 A. Further development of the project was passed on to Blohm & Voss.

**Bf 109 T-2 Specifications**

**Dimensions**

| | | |
|---|---|---|
| • Assembled | Wingspan | 11 080 mm |
| | Length | 8 785 mm |
| | Height (tail down) | 2 600 mm |
| | Wing area | 17.5 $m^2$ |
| • Wings Removed | Maximum span | 2 115 mm at the wheels |
| | Span of folded wings | 4 590 mm |
| | Span of horizontal stab. | 3 000 mm |
| | Height (tail down) | 2 600 mm |
| **Power Plant** | 1 DB 601 N, 1,175 H.P. at 2,600 rpm (C 3 fuel) | |
| **Propeller** | 3-blade VDM variable-pitch propeller, diameter 3 000 mm | |
| **Armament** | 2 synchronized MG 17s with 1,000 rounds per gun<br>2 wing-mounted MG FF with 60 rounds per gun | |
| **Maximum Allowable Speed** (indicated) | Level flight at ground level | 480 kph |
| | Flaps fully extended | 250 kph |
| | Undercarriage extended | 350 kph |
| | Dive from 0 to 3 000 m | 750 kph |
| | 3 000 to 4 000 m | 730 kph |
| | 4 000 to 5 000 m | 690 kph |
| | 5 000 to 6 000 m | 650 kph |
| | above 6 000 m | 510 kph |

The Bf 109 T-2 is not equipped for catapult launches or arrested landings.

**Table of Loads Bf 109 T-2**

| | Weight in kilograms | | | |
|---|---|---|---|---|
| Stress group | H5 | H4 | H4 | H4 |
| Purpose case | A | B | C | D |
| | normal | overload | overload | overload |
| **Empty Weight** | 2 000 | 2 000 | 2 000 | 2 000 |
| Additional equipment | 225 | 253 | 264 | 232 |
| Equipped Weight | 2 225 | 2 253 | 2 264 | 2 232 |
| Fuel 400 l | 312 | 312 | 312 | 312 |
| Fuel 300 l + drop tank | 252 | | | |
| Oil 29.5 l | 27 | 27 | 27 | 27 |
| Oil 9 l (with drop tank) | 8 | | | |
| Pilot + parachute + special clothing | 100 | 100 | 100 | 100 |
| Ammunition for fuselage MG 17s | 59 | 59 | 59 | 59 |
| Ammunition for wing MG FFs | 40 | 40 | 40 | 40 |
| Bombs 1 x 250 or 4 x 50 kg | 250 | 100 | | |
| Armor plate | 24 | 24 | 24 | 24 |
| Head armor | 13 | 13 | 13 | 13 |
| **Takeoff Weight** | 2 800 | 3 780 | 2 939 | 3 067 |

## Water Landings with the Bf 109 Experiments with Models

The carrier aircraft was employed as a land aircraft over water; therefore, the possibility of a forced landing in the water had to be taken into consideration. Land-based aircraft with landing speeds under approximately 70 kph had ditched with relatively little damage (Ar 66, He 63). However, as landing speed increased, so too did the tendency to overturn, especially with aircraft with fixed undercarriages. It therefore seemed advisable to employ models to in-

**Maximum Allowable Takeoff Weight**

| | | |
|---|---|---|
| Normal | Case A in H5 | 2 800 kg |
| Overload | Case B in H4 | 3 078 kg |
| Overload | Case C in H4 | 2 939 kg |
| Overload | Case D in H4 | 3 067 kg |

**Performance**

a) Climb

| Height km | Rate of climb m/sec | Time to climb min. | RPM atm | Boost Pressure climb | Speed at Best |
|---|---|---|---|---|---|
| 0 | 17 | 0 | 2,400 | 1.25 | 250 kph |
| 2 | 17 | 2 | 2,400 | 1.25 | 230 kph |
| 4 | 17 | 4 | 2,400 | 1.25 | 210 kph |
| 5 | 16 | 5 | 2,400 | 1.20 | 200 kph |
| 6 | 13 | 6.4 | 2,400 | 1.06 | 195 kph |
| 8 | 7 | 10.0 | 2,400 | 0.84 | 180 kph |
| 9 | 4 | 12.5 | 2,400 | 0.75 | 175 kph |
| 10 | 1.5 | 17.4 | 2,400 | 0.68 | 170 kph |

b) Horizontal Flight:

| Height km | Cruise 2,300 rpm | | Full Power 2,400 rpm | | Takeoff and Emergency Power 2,600 rpm | |
|---|---|---|---|---|---|---|
| | boost (atm) | speed (kph) | boost (atm) | speed (kph) | boost (atm) | speed (kph) |
| 0 | 1.15 | 433 | 1.25 | 475 | 1.35 | 490 |
| 2 | 1.15 | 488 | 1.25 | 515 | 1.35 | 528 |
| 3 | 1.15 | 510 | 1.25 | 535 | 1.35 | 546 |
| 4 | 1.15 | 532 | 1.25 | 555 | 1.35 | 565 |
| 5 | 1.15 | 552 | 1.25 | 570 | | |
| 6 | 1.08 | 552 | 1.25 | 570 | | |
| 8 | 0.85 | 535 | 0.88 | 560 | | |
| 9 | 0.76 | 544 | 0.78 | 544 | | |
| 10 | 0.68 | 515 | 0.70 | 520 | | |

c) Takeoff and landing at all-up weight of 2 800 kg:

| | |
|---|---|
| Takeoff distance to height of 20 m | = 500 m |
| Liftoff speed | = 120 kph |
| Landing distance from height of 20 m | = 700 m (non-arrested) |
| Landing speed | = 130 kph |

**Range:**

| Altitude atm | Engine Boost Output | Engine Pressure | Speed RPM | TAS kph | Flying Time hrs. | Distance km |
|---|---|---|---|---|---|---|
| 1 | Cont. | 1.15 | 2,300 | 460 | 1 hr 10 min | 545 |
| | L.g.R. | 0.75 | 1,300 | 320 | 2 hr 50 min | 915 |
| 3 | Cont. | 1.15 | 2,300 | 485 | 1 hr 10 min | 555 |
| | L.g.R. | 0.75 | 1,300 | 330 | 2 hr 40 min | 865 |
| 5 | Cont. | 1.15 | 2,300 | 550 | 1 hr 10 min | 585 |
| | L.g.R. | 0.80 | 1,400 | 355 | 2 hr 10 min | 740 |
| 7 | Cont. | 1.0 | 2,300 | 545 | 1 hr 15 min | 595 |
| | L.g.R. | 0.70 | 1,600 | 375 | 2 hr 05 min | 730 |
| 9 | Cont. | 0.80 | 2,300 | 520 | 1 hr 25 min | 615 |
| | L.g.R. | 0.65 | 1,900 | 455 | 1 hr 50 min | 730 |

vestigate the ditching characteristics of faster aircraft (Bf 109) in order to avoid unnecessary losses in human life and materiel. The model experiments took place from 22 September to 23 October, and on 22 December 1937 in the test facilities of the DVL's Institute for Marine Flight in Hamburg. The object of the tests was to identify unsatisfactory landing characteristics and improve them through modifications to the aircraft. At the same time efforts were made to determine the most favorable touchdown angle and landing speed for a forced landing on the water. The Bf 109 Es

Sketch of model experiments for forced landings on water in the test facility of the Institute for Marine Aviation in Hamburg.

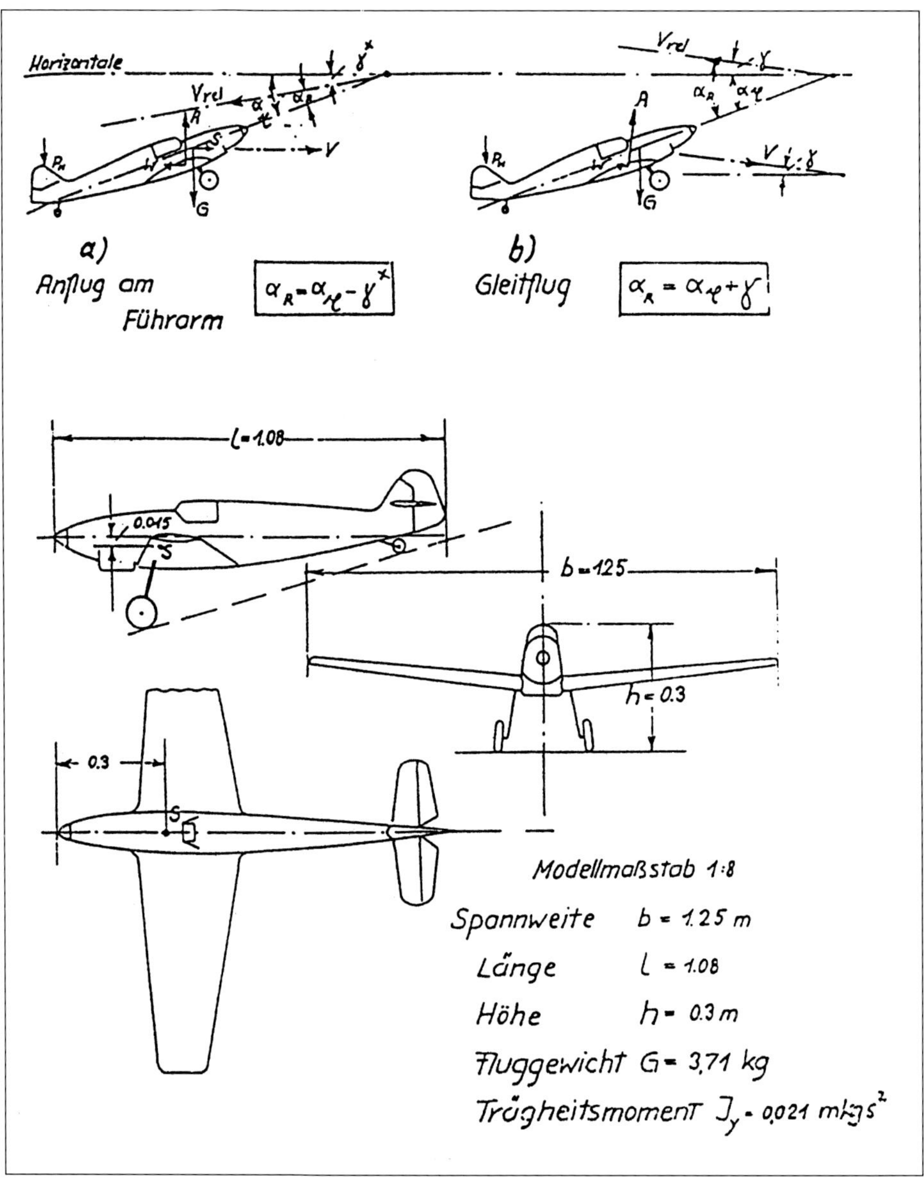


model was produced in 1 : 8 scale, which resulted in a wingspan of 1,250 mm, a length of 1,080 mm and a height of 30 mm (with extended undercarriage). The model weighed 3.71 kg. The evaluation of the numerous diagrams will not be discussed here. All activities were carried out with a view toward the coming aircraft carrier *Graf Zeppelin* and the Bf 109 T carrier fighter.

# Various Experiments Involving the Bf 109

Telling the entire story of the program of experiments and tests involving the Bf 109 E would exceed the scope of this book. Some of the tests have already been mentioned in describing the various prototypes and versions. The following section will describe some of the lesser-known developments.

### Aircraft Towing with the Bf 109 for Increased Range

Carried out before trials with the jettisonable fuel tank, its purpose was to increase the aircraft's relatively limited endurance in action and thus its range. The experiments were carried out in conjunction with "Operation Sea Lion" (the German invasion of Great Britain).

Almost every publication on the Bf 109 contains the in-flight photograph of the Bf 109 E-3 *Werknummer* 1953, CE + BM. In every case the photo is interpreted incorrectly, as the "barrel" in the center of the propeller spinner is identified as belonging to a MG C/30 L engine-mounted cannon. In reality it was a tow coupling, to which a tow cable was attached.

After several release experiments the tow flight took place on 18 April 1940. The tug aircraft, Bf 110 C-1 WerkNr. 959, CE + BU, was flown by Dr. Würster. In the cockpit of the Bf 109 was Heini Dittmar. After taking off from Augsburg the combination flew to Munich-Riem (10:58 to 11:10 AM). The return flight took place from 6:15 to 6:31 PM. Dittmar received a bonus of 2,000 *Reichsmark* for the flight. Messerschmitt abandoned this rather unusual method of towing because the easier to use drop tank was introduced into service.

During takeoff and flight the propeller was set to "Glide Position" and the engine remained at idle. Once released, the Bf 109 was able to carry out a normal landing with propeller and engine speed at standard settings.

### Increased Engine Output through Use of the GM 1 System

A partial description may be found in the chapter on the Bf 109 E-7/Z. The first system with pressure-liquefied nitrous-oxide (GM 1 fluid) was installed in Bf 109 E-7 *Werknummer* 3809 (GA + LX). In summer 1941 the *E-Stelle Rechlin* made a thorough evaluation of the system at altitudes of 8 to 11 kilometers. Flight tests were preceded by bench runs, which revealed that the greatest power gain with GM 1 fluid was achieved by injecting it prior to the supercharger.

**The tow coupling is clearly visible in this close-up photograph; in many publications it is erroneously identified as the barrel of an engine-mounted cannon.**

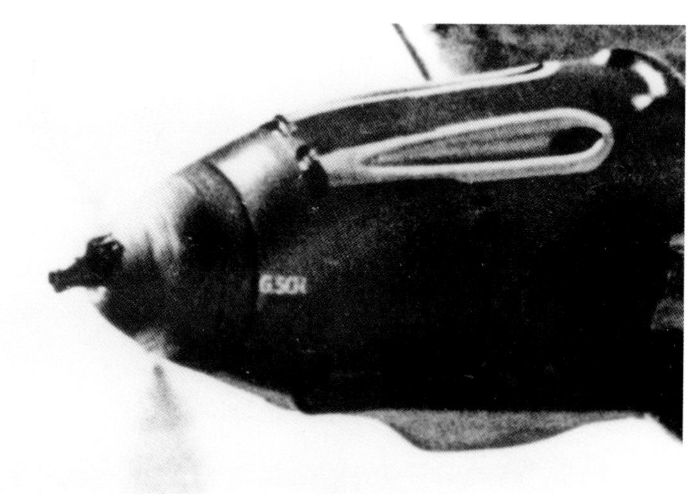

The horizontal flights at altitudes of 8 to 11 kilometers were conducted as full-throttle flights at 2,400 rpm. The maximum speed was achieved approximately 1 to 2 minutes after injection of the GM 1 fluid was initiated. Output fell slowly after the third minute, returning to normal after the fifth minute (when the liquefied gas was expended). Engine output using the system was 1,500 H.P. A capacity of 30 kg of GM 1 fluid and an average flow rate of 100 g/sec provided five minutes flying time. The increase in true airspeed lay between 60 and 105 kph (depending on altitude).

The system was only supposed to be used when really needed (air combat, emergencies). It proved a success, and from 1941 was employed by a growing number of fighters and destroyers (with DB and Jumo engines).

**Bf 109 V21, WerkNr. 1770, D-IFKQ, with Radial Engine**

At the direction of the GL/C-E3, in 1938 a comparison was to be made between the installation of radial and inline engines in a modern single-seat fighter. The purposes of this experiment included evaluation of the aerodynamics, handling characteristics, view while taxiing, etc. The BMW 139 and BMW 801 were the selected radial engines.

On 18 August 1939 the Messerschmitt AG of Augsburg received a contract (LC 6 IIa Nr. 47 1b) to build a test-bed powered by a twin-row radial engine. BMW-Munich had only begun the design of the

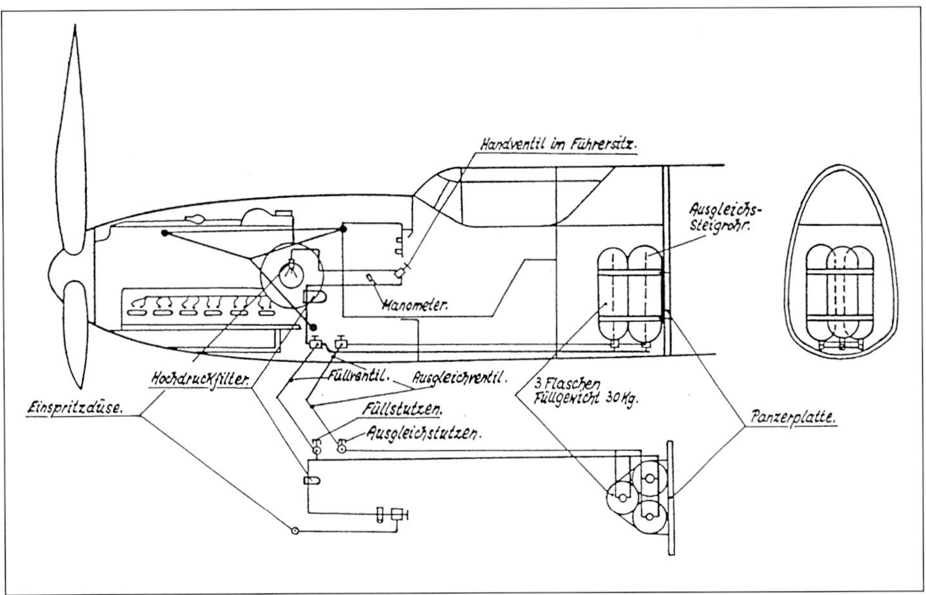


Drawing of the GM 1 power boost system in the Bf 109 E.

Bf 109 E-3 with tow coupling in the propeller spinner, used for range increasing experiments. The tow experiments with the Bf 109 E-3 were linked to "Operation Sea Lion," the planned invasion of England.

BMW 801 (successor to the BMW 139) in autumn 1938, consequently, an American engine, the Pratt and Whitney Twin Wasp (twin-row radial engine producing 1,200 H.P. for takeoff), had to be installed in the V21. The value of the contract was 64,983.59 *Reichsmark*. By 31 March 1939 the V21 was 95% complete.

The conversion process included widening the fuselage forward of Frame 7 to accept the radial engine, which was circular in cross section, redesign of the cockpit canopy as a bubble-type with sliding hood (similar to that later used by the Fw 190). The engine mount was also a completely new design. Changes had to be made to the wing-fuselage junction. Date of the aircraft's first flight is not known (mid-1939?). In any case, Dr. Würster and Fritz Wendel flew the V21 at Augsburg on 18 August 1939. Further test flights followed, the last entry in Wendel's log book appearing on 4 September 1939 (operating reliability flight). The machine then went—probably by way of Rechlin—to the DVL's flight department, which had been moved to Braunschweig-Völkenrode at the start of the war, where it was assigned the code KB + II. It was used to familiarize aircraft designers in training. Flights are known to have been made as late as September 1940. Fritz Wendel made the first flight in the BMW 801-powered (Type A/0, WerkNr. 80153) Bf 109 X (WerkNr. 5608, D-ITXP) on 2 September 1940.

The purpose of the installation of a Twin Wasp in the Bf 109 V21 was to compare the aerodynamics, flight characteristics and view from the cockpit during taxiing of machines equipped with radial and inline engines.

## Bf 109 B-1, WerkNr. 1013, D-IMSY, with Skis

Before D-IMSY was released to be scrapped (it had been "on loan" to Messerschmitt) it served as a prototype for a snow-ski installation (mainwheels and tailwheel, device no. 8-3903/13). The skis, which were provided by the Hugo Heine Company, were rigid boot skis mounted on shock absorbers about their longitudinal and lateral axes. The skis were fitted at Augsburg in the summer of 1940.

For its tests, which took place in Gardermoen, Norway, from mid-December 1940 until the end of January 1941, the *E-Stelle Rechlin* used a Bf 109 E. Trials had to be broken off after the 87th flight as the left ski tilted down and to the left after takeoff, making a landing impossible. The pilot abandoned the machine, which was destroyed in the subsequent crash and explosion. It was recommended that the trials be resumed with modified skis.

The E-*Stelle*'s findings: takeoff and landing normal, stability about the lateral axis had become very weak, however, the aircraft remained fully operational for its specialized role. The skis were found to be far from fit for service use.

The *E-Stelle* resumed trials with newly-developed skis mounted on Bf 109 F WerkNr. 8195 (VD + AJ) at Dorpat (Tartu, Estonia).

## Bf 109 V1, WerkNr. 758, D-IABI Roll-off and Spin experiments

Roll-off is a sudden, strong movement about the longitudinal axis. This phenomenon, whether intentional or unintentional, is usually preceded by an excessively sharp pull-up and degradation of airflow. Roll-off could also be induced by irregular extension of the slats. The experiments to discover the cost were very expensive (ca. 2,000,000 Reichsmark). They were begun by BFW in Augsburg and continued by the DVL at Berlin-Adlershof. The tests lasted from August 1936 until the beginning of

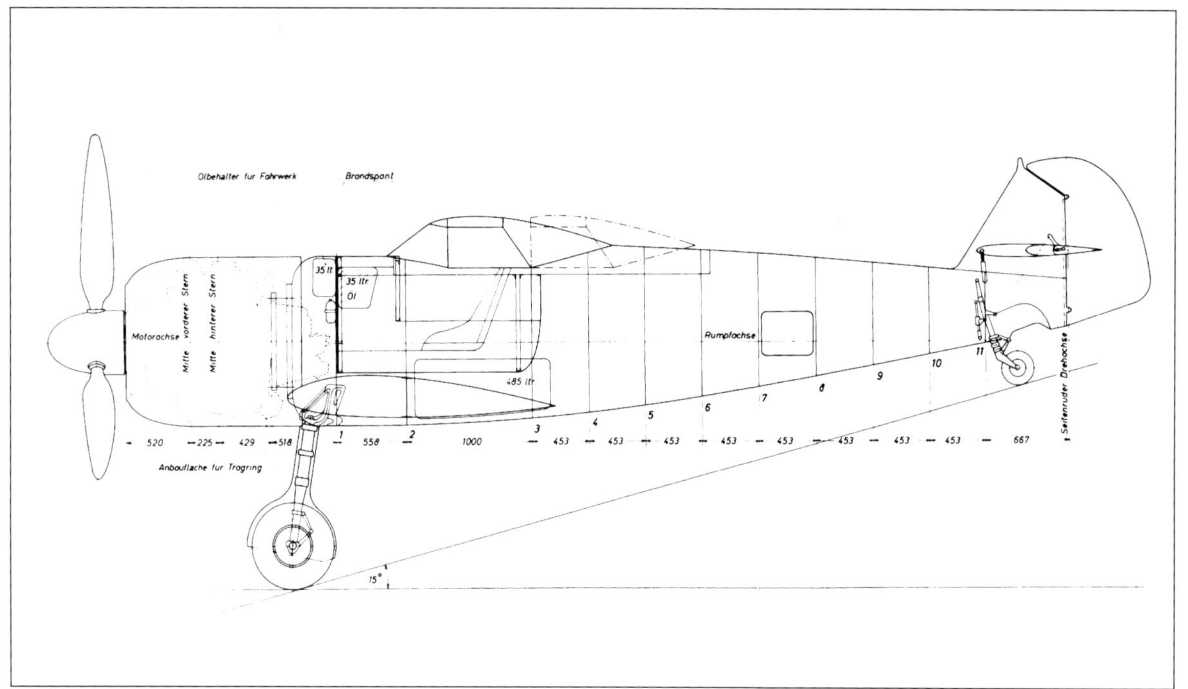


Side view of the Bf 109 X with BMW 801 engine (above, drawing by Mohr). A Bf 109 E with snow skis was tested in Norway from December 1940 to January 1941 (below).

1941. Used for the experiments dubbed "Roll-offs and spins with free and shortened slats" was the Bf 109 V1, which had returned to Augsburg from Travemünde on 7 July 1936. Various experimental flights conducted by the *E-Stelle Travemünde* and the *E-Stelle Rechlin* showed that roll-off behavior could be improved in loops and turns by freeing the slats even when the landing flaps were fully retracted. Previously the slats had been locked in position flush with the wing and were not released until the landing flaps had been lowered roughly ten degrees or more.

The first two flights with retracted flaps and slats "free" and "locked" during roll-off were undertaken by Dr. Würster on 24 August 1936. Wool tufts had been affixed to the entire upper surface of the wing, and their movements were photographed during flight by a 16-mm movie camera installed in the cockpit.

Dr. Würster conducted four more flights on 26 August 1936. His mission was to obtain roll-off photos during right and left turns with the slats "free." The flight tests revealed that during aerobatics and turns the Bf 109's roll-off behavior was

significantly better with the slats unlocked and landing flaps retracted than with the slats locked flush. One interesting note is the extensive correspondence that BFW maintained with the Handley Page company (Mr. Lachmann, London) in spite of the military nature of the project.

At his suggestion a compressed air retraction system for the slats was installed in the wing, this being intended for dangerous spin situations. Another thirteen test flights were carried out from 23 September to 19 December 1936, of which flights four to eight were conducted by Kurt Jodlbauer. On 24 September 1936 the Bf 109 V1 received a new Rolls Royce Kestrel II S for further trials. The retraction mechanism was never used in any of the spin tests, as the spin situation never became "dangerous."

A second series of tests (five flights) was supposed to determine whether the Bf 109's "pleasant" roll-off and spin characteristics could be influenced by shortening the slats. The planned installation of wing-mounted weapons required a shortening of the inner slats by 635 mm in order to keep the blast tubes clear of the slat area. Wind tunnel trials by the AVA Göttingen suggested that improved spin characteristics could

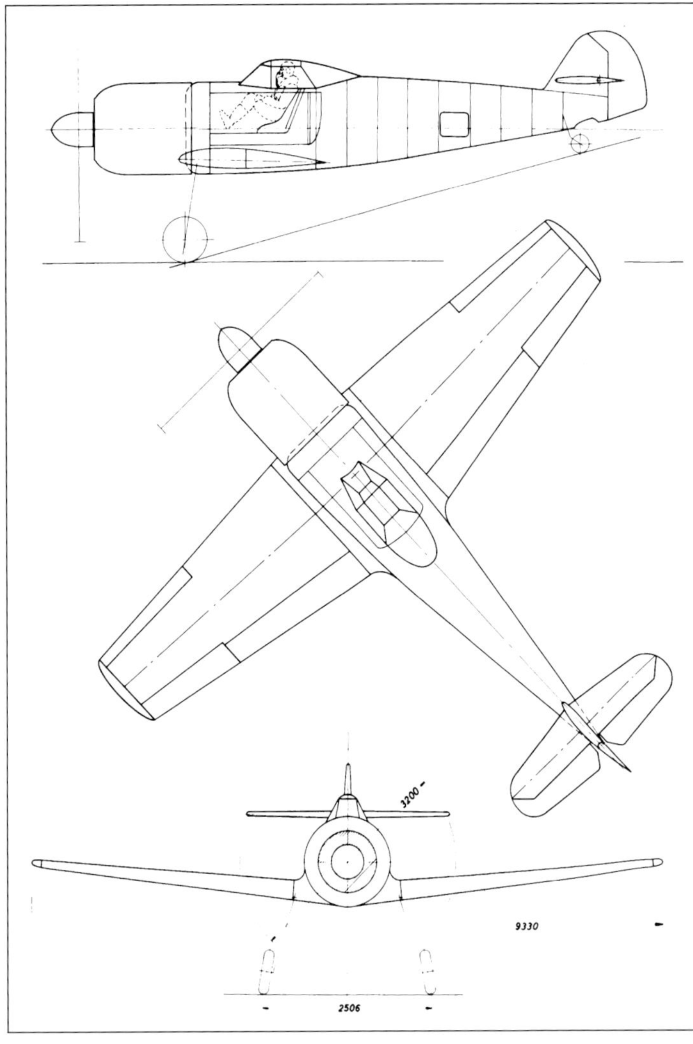

**Three-view drawing of the Bf 109 X with BMW 801 (above, drawing by Mohr). The ski trials in Norway were broken off after the 87th flight (right).**

The skis for the Bf 109 E were delivered by the Heine Company and were fitted on shock-absorbing mounts.

be expected as a result of shortening the slats.

The first flight by the Bf 109 V1 took place on 9 January 1937 in the hands of Dr. Würster, the second on 11 January with the same pilot at the controls. The third flight on 13 January was made by Dr. K. Jodlbauer. Bf 109 B-1 WerkNr. 1013, D-IMSY, was used for the fourth and fifth flights by Dr. Würster on 26 June and 3 July 1937. The result of all the flights was that the shortening of the leading edge slats did not have any serious negative effects.

### Bf 109 B-1, WerkNr. 290, D-IHHB Wing Fence

The machine (built under license by Erla) arrived at Augsburg on 3 July 1937 for continuation of experimental flights. In the prototype shop it was converted into a test aircraft specifically for roll-off experiments. A wooden wing with three degrees of wash-in but no slats was installed. The wing did not differ from the standard 109 wing in outline or profile. A ducted mast was installed in front of the cockpit, and a faired cine camera was mounted on the fin to film the airflow movements of the wool tufts glued to the wing.

For the 83rd flight (16 September 1939, Fritz Wendel) a so-called boundary fence was installed 300 mm inside the outer end of the landing flap, the first time such a device had been used in the history of aerodynamics. The fence was 930 mm long, 300 mm high in front and 70 mm high at the back. The purpose of the fence was to halt the spanwise flow during roll-off and delay the collapse of lift.

These fences were first used on a production aircraft in 1947 on the Russian MiG 15 fighter. They were later used on the British Comet and French Caravelle airliners. The wooden wing failed to display the expected characteristics. Consequently, the DVL in Berlin-Adlershof took over the subsequent execution of the program with D-IHHB, which was equipped with various installations such as hinged spoilers, disruptive surfaces (rough upper surfaces) and boundary fences of various sizes. At the beginning of 1940 measurements confirmed the effectiveness of the boundary fences on a Bf 109 E, however, Messerschmitt saw no direct requirement to apply these to the Bf 109. As before, the leading edge slat remained the most effective means. The invention of the boundary layer fence went back to Messerschmitt's aerodynamic department. Scientific investigation was done

**In preparation for a series of tests involving boundary layer fences this Bf 109 B-1 was fitted with a yaw measuring device mounted on a mast in front of the cockpit and a camera on the vertical fin (above). The Bf 109 V24 was tested in the wind tunnel in Chalais Meudon near Paris (right).**

by Dr.-Ing. Kurt Liebe. A patent application was submitted (No. 700625) and one was granted to DVL in 1944.

### Bf 109 B-1, WerkNr. 1014, Maximum Allowable Diving Speed

On behalf of BFW Augsburg company pilot Dr. Kurt Jodlbauer was to determine the terminal velocity of the Bf 109 at Rechlin. The tragic outcome of these tests, which followed several successful dives, is revealed in the *Technische Amt*'s daily reports Nos. 208 and 209:

"On 17 July 1937 a Bf 109 flown by company pilot Dr. Jodlbauer crashed into the Müritzsee near Vietzen/Rechlin, killing the pilot. The pilot had undertaken diving flight tests from 5 000 meters."

"Probable cause of the crash: the aircraft had been trimmed so nose-heavy that Jodlbauer did not have the strength to level out using the elevators and the horizontal stabilizer incidence control. The previous evening Jodlbauer had stated that he would have to increase the amount of nose-down trim in order to prevent the aircraft from leveling out of the planned dive sooner than desired. There is no evidence of structural failure of the aircraft or aircraft components."

Technical specifications and pilot's handling notes listed the Bf 109's maximum allowable diving speed as 750 kph. At the beginning of the war, however, the units reported difficulties (movements about the normal axis) in dives and ensuing accidents not due to enemy action, most of them fatal. It is probable that the pilots had far exceeded the maximum allowable diving speed in air combat.

For this reason *Flugbaumeister* Dipl.-Ing. Heinrich Beauvais (*E-Stelle Rechlin*) undertook preliminary experiments, however, these were not recorded. In January 1941 he exceeded the 750 kph mark, and only after great physical exertion was he able to level out the aircraft at an altitude of approximately 4,000 meters. Further tests of this type were temporarily forbidden. But the problem was still awaiting a solution. Not until 12 March 1943, six years after the crash of Dr. Jodlbauer, did Messerschmitt test pilot Likas Schmid reach the Bf 109's maximum diving speed of 906 kph at an altitude of 5.8 km (after a maximum power dive from 10,700 meters). He achieved this feat while flying Bf 109 F WerkNr. 9228, TH + TF (with G wings, larger G tail surfaces and ejector seat).

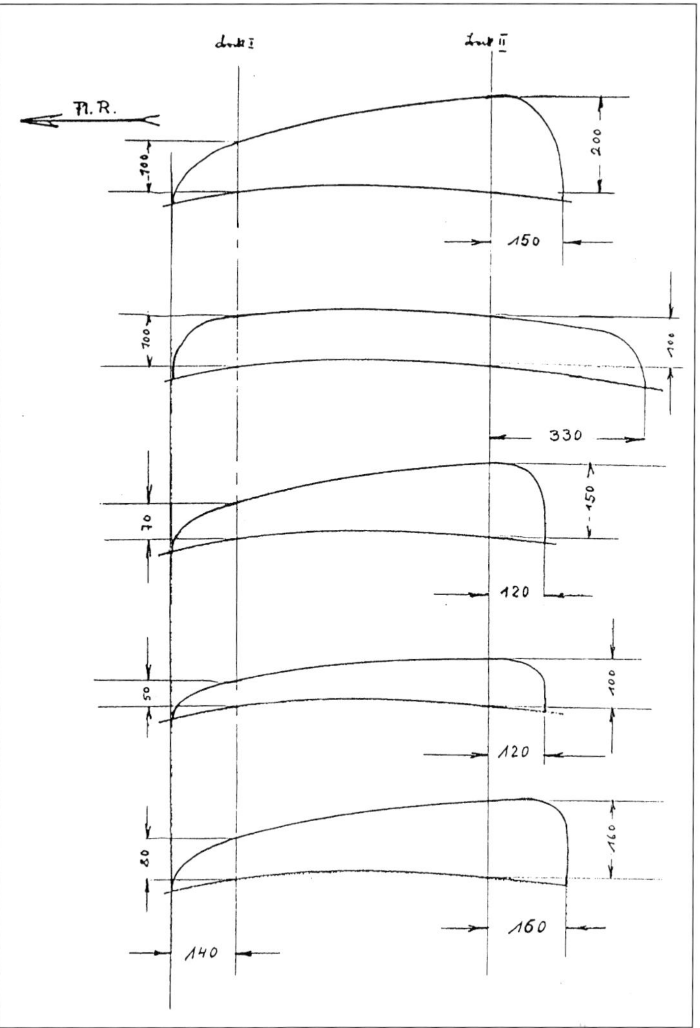


Dr. Kurt Liebe proposed five different boundary layer fences for the Bf 109.

A Bf 109 E-3 in the big wind tunnel of the Hermann Göring Aviation Research Institute in Völkenrode near Brunswick.

## Bf 109 V24, WerkNr. 1929, D-ITDH (from April 1940 CE + BH) Drag and Lift Measurements

The measurements took place in mid-1941 in the Chalais-Meudon wind tunnel near Paris. The first flight of the V24 took place in February 1939. It was used for intensive testing at Augsburg in 1939 and 1940. Most of these flights were carried out by Fritz Wendel and Dr. Würster. In March 1942 work began to prepare the machine for wind tunnel measurements. The machine was delivered to Paris with no propeller, an F-type spinner—the propeller blade openings were sealed over—and a sealed supercharger air intake (hood placed over opening). The first measurements were carried out with the V24 in "normal" condition, which was worse than the production standard at that time, as it had seen much use as a test aircraft and had often been converted. There then followed another series of measurements with the machine in "improved" condition, which was achieved through better surface finish (filling, sanding and polishing), modified radiator fairings and undercarriage wheel well doors. The entire aircraft was placed in the center of the air stream and was borne on a carriage with three supports.

The improvements resulted in a reduction in drag which was equivalent to an increase of 44 kph over the original speed of 600 kph. No further details of the measurements will be provided here.

From 1 October 1941 the measurements were continued in the AVA's wind tunnel in Göttingen. On 1 February 1942 the machine returned to Augsburg, where a central radiator and oil cooler were tested. This was part of the preliminary trials for the Me 309.

## Over-Wing Containers for the Bf 109

As early as 1942 the proposal was made by the FGZ[1] to provide increased capacity of existing types (He 111, Ju 87, Bf 109) through the use of easily installed and removed over-wing cabins ("riders"). Based on the first proposals this space was to be used to transport men and equipment.

The first flying mock-ups were built at the beginning of 1943, for the Klemm Kl 25. Experimental flights were made to determine their effect on flight characteristics. This was followed in the summer of 1943 by the design and construction of jet-

[1] Graf Zeppelin Research Institute, Stuttgart-Ruit.

**Bf 109 E-3 vor dem Tor des Windkanals in Völkenrode.**

tisonable, over-wing cabins for the Bf 109, each of which accommodated one man (ground crew). The form selected for these preliminary trials had no unfavorable effect on flight characteristics and produced only a small reduction in airspeed. The passenger was accommodated in a prone position wearing a seat-type parachute, and a small teardrop-shaped window was placed in the front quarter of the rider. The two riders sat far enough away from the fuselage that they did not have any affect on the horizontal tail. The following table shows the capacity and resulting speed loss for the empty mock-up for the Bf 109 (Kl 25 and Ju 87 provided for comparison).

Speed Loss for the Bf 109

| Aircraft Type | Purpose-Volume $m^3$ | Speed Loss |
|---|---|---|
| Kl 25 | Preliminary trials 3.24 | 25 kph from 150 to 125 kph |
| Ju 87 | Personnel transport 5.72 | 38 kph from 318 to 280 kph |
| Bf 109 | Personnel transport 0.86 | 21 kph from 410 to 389 kph |

The experiments were flown with a Bf 109 E.
On the basis of the results of wind tunnel and flight experiments, especially with the Bf 109's riders, the FGZ proposed over-wing fuel tanks of a similar shape. These were tested on three Fw 190s.

# Propeller Tests

**The MeP 8 was the largest variable-pitch propeller developed by Messerschmitt. It was intended for engines in the 1,400 to 2,000 H.P. range.**

At this point we turn to a theme that has received scant mention in publications dealing with the Messerschmitt company and its aircraft, specifically Messerschmitt work on propellers. The term propeller is not entirely accurate in this case, because it was not propeller blades, but instead variable-pitch propeller systems which Messerschmitt designed, and in some cases built. One of these, the Me P 6, was tested on several Bf 109 Es, including at the *E-Stellen Rechlin* and *Travemünde*.

Professor Messerschmitt wanted to pursue his own projects independent of the RLM, and for this purpose in mid-1936 he founded the "Preparation Bureau" (VB) within the BFW-Werke in Augsburg. This was later renamed "Special Development" (SE). In charge of this department was Dipl.-Ing. Robert Prause. Its goal was to increase the effectiveness of existing propellers for the Bf 109 and future Messerschmitt developments. Ideally suited to this was Prof. Messerschmitt's principle of the MeP 6 for rapid pitch change, for which he applied for a patent. This allowed the propeller to be used as an effective landing and dive brake. Using this type of "reverse thrust" reduced the Bf 109's landing roll to almost 120 meters. Furthermore, to the amazement of the watching experts, the aircraft could be taxied backwards with the engine running at high speed.

**In der numerischen Reihenfolge kam es zu folgenden Entwicklungen:**

| | |
|---|---|
| MeP 2 | 2-Blatt-Verstellschraube für Triebwerke bis 200 PS |
| MeP 6 | 3-Blatt-Verstellschraube für Triebwerke von 600 – 1000 PS |
| MeP 7 | 2-Blatt-Verstellschraube für Triebwerke von 200 – 600 PS |
| MeP 8 | 3-Blatt-Verstellschraube für Triebwerke bis 2000 PS (DB 603 G/H, Jumo 213) |
| MeP 10 | 3-Blatt-Verstellschraube für Triebwerke Planung, für hydraulische Verstellung |
| MeP 12 | 6-Blatt-Verstellschraube für Triebwerke 2 x MeP 6, gegenläufig. |

**Vorgesehene Verwendung in:**

| | |
|---|---|
| MeP 2 | Bü 181 |
| MeP 6 | Bf 109, Bf 110, Me 210/410, Me 264, Ju 87, Do 24, BV 138, Ar 196 |
| MeP 7 | Bf 108, Fh 104, Si 204, Ar 79, Ar 96 B |
| MeP 8 | verschiedene mit DB 603 oder Jumo 213. |

## Chronological Development of the MeP 7

1936 saw the development of a privately-financed two-blade variable-pitch propeller, primarily for the Bf 108 *Taifun*, which received the designation MeP 7. The first test installation (Schwarz blades, 2,350 mm diameter) took place in mid-1937 using Bf 108 WerkNr. 990, D-IRNU. The first flight was undertaken by the head of the VB, R. Prause, on 8 July 1937. The subsequent testing may have gone on until April 1938.

Elly Beinhorn made her East-Asian flight in Bf 108 WerkNr. 2019, D-IROS, from 20 April to 30 June 1939; it also served as an endurance test for the MeP 7, which proved a success.

The Bf 108 was built in the Regensburg factory, where it was picked up by the owner herself. After this flight the MeP 7 was cleared for production and became a standard feature of the Bf 108.

Blade position was set mechanically by means of a hand crank on the instrument panel. Surviving records do not make it possible to state with certainty how many Bf 108s in total were equipped with the

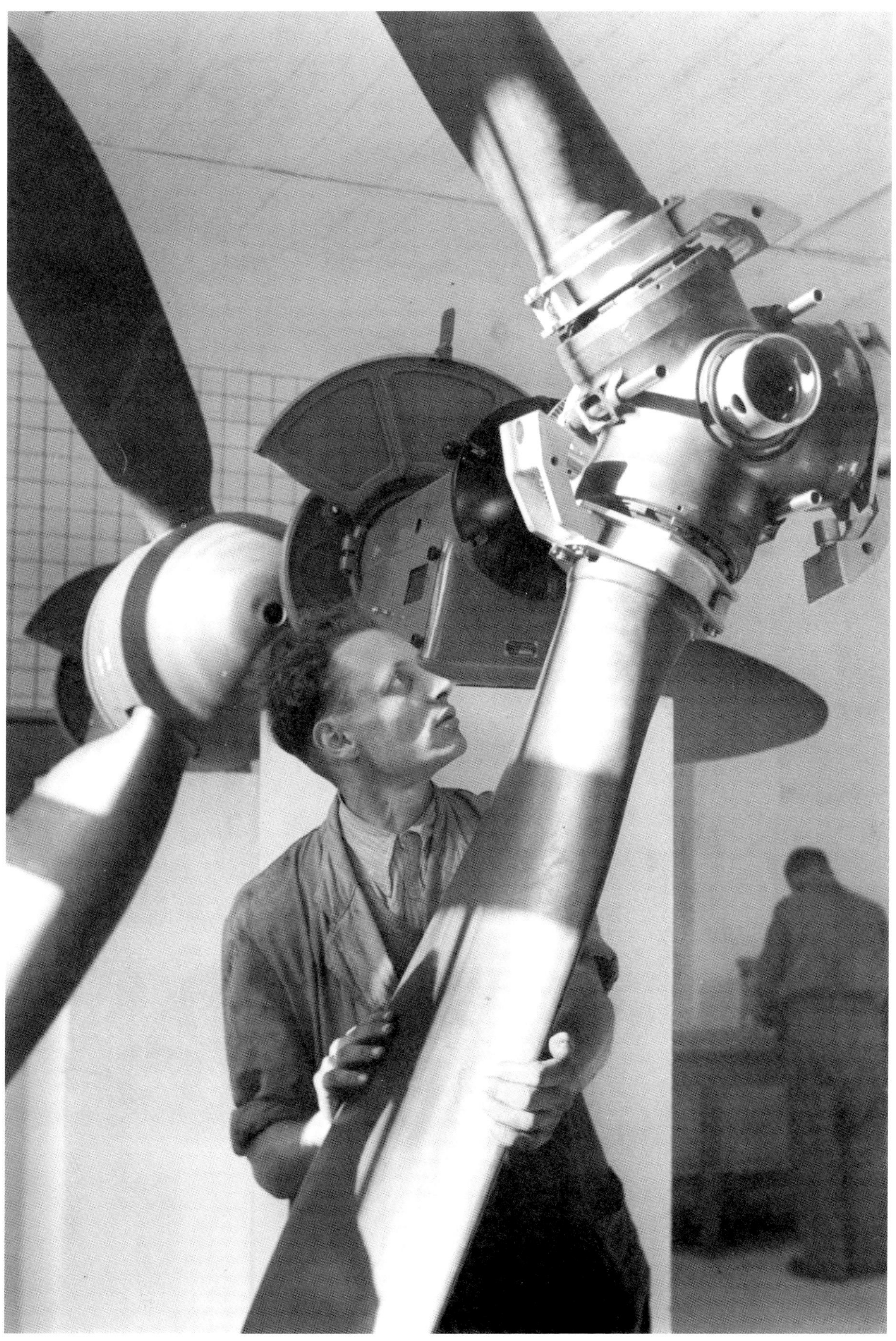

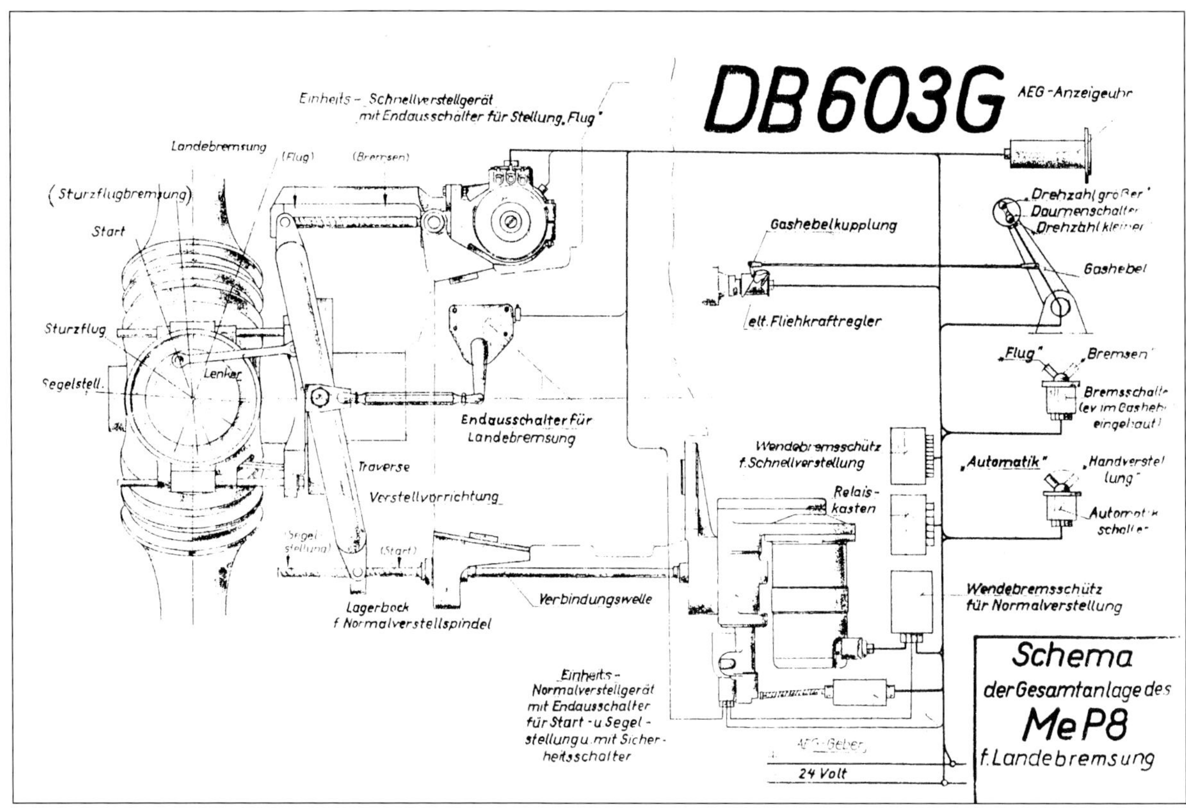


Schematic drawing of the Messerschmitt MeP 8 propeller pitch control system with possible use as a landing brake.

MeP 7. In any case, it was the only Messerschmitt propeller to be built in significant numbers.

In April 1940 an experimental installation with automatic pitch control was made in Bf 108 B-1 WerkNr. 2122, D-IOIO (from Sept. 1941 DI + CB). The machine remained in use by Messerschmitt as a courier aircraft until January 1942. As well, a small number of MeP 7 S propellers (S = gliding position) were built for the As 410, As 411 and HM 508 power plants. Messerschmitt tested the propeller on the Ar 96 B WerkNr. 1259, GJ + YY.

## MeP 6

By mid-1937, one year after the initial development of the MeP 7, the VB had already designed a variable-pitch propeller for higher performance engines and initiated construction of a prototype. It could be used for two-, three- or four-blade propellers. The initial version was installed in the Bf 109 V10a WerkNr. 1010, D-IAKO, which was powered by a Jumo 210 D. Static thrust measurements over the entire blade angle range of 360° revealed that it was possible to achieve a reduction in land-

**Versuchsträger für den MeP 6**

| | Werk-Nr. | Kennzeichen | Bemerkung |
|---|---|---|---|
| **Bf 109 E-3** | 1797 | CE+BL | Vergleich Ju-, VDM- und MeP 6 Verstellschrauben |
| **Bf 109 E-7** | 1946 | GH+NU | |
| **Bf 109 E-7** | 1952 | CE+BM | Dauererprobung (500 Stunden) |
| **Bf 109 E** | 1995 | GJ+AY | |
| **Bf 109 E** | 4950 | VG+CX | Juni 1941 in Rechlin |
| **Bf 109 V23** | | 1801 | D-ISHN CE+BG, 15. Juli 1941 Bruch |

ing roll distance through propeller braking and that it could also be used as a dive brake. In order to achieve this, however, the speed of pitch change had to be increased from 8 to 9 degrees per second to 40 to 50 degrees per second. This made it possible to reduce the Bf 109's landing roll from 630 meters to approximately 170 to 120 meters. An 800-watt pitch motor was anticipated for switching the blades to negative pitch angles for braked landings; this was capable of achieving pitch change speeds of 60 degrees per second (360 degrees in 6 sec.).

The good results achieved in flight trials to date led the RLM to get involved, and on 21 January 1938 it issued contract LC IV 2a B.Nr.12.13 Nr. 124/37. In doing so it assumed the costs of this and other landing brake experiments up to 32,845 Reichsmark. In addition, several experimental versions of the MeP 6 were ordered for two- and three-blade power propellers. The first twin-engined test-bed for the use of reverse pitch as a dive brake was Bf 110 B-1 WerkNr. 918, D-ADVO. It was equipped with two Jumo 210 G engines driving MeP 6 propellers.

The trials took place between 28 October 1938 and 7 September 1939. At the request of the RLM the MeP 6 was now installed in various aircraft types for test purposes.

The following Bf 109 Es are known to have been used as test-beds for the MeP 6 (list incomplete), some of which flew with the E-St*ellen Rechlin* and *Travemünde* (see table below).

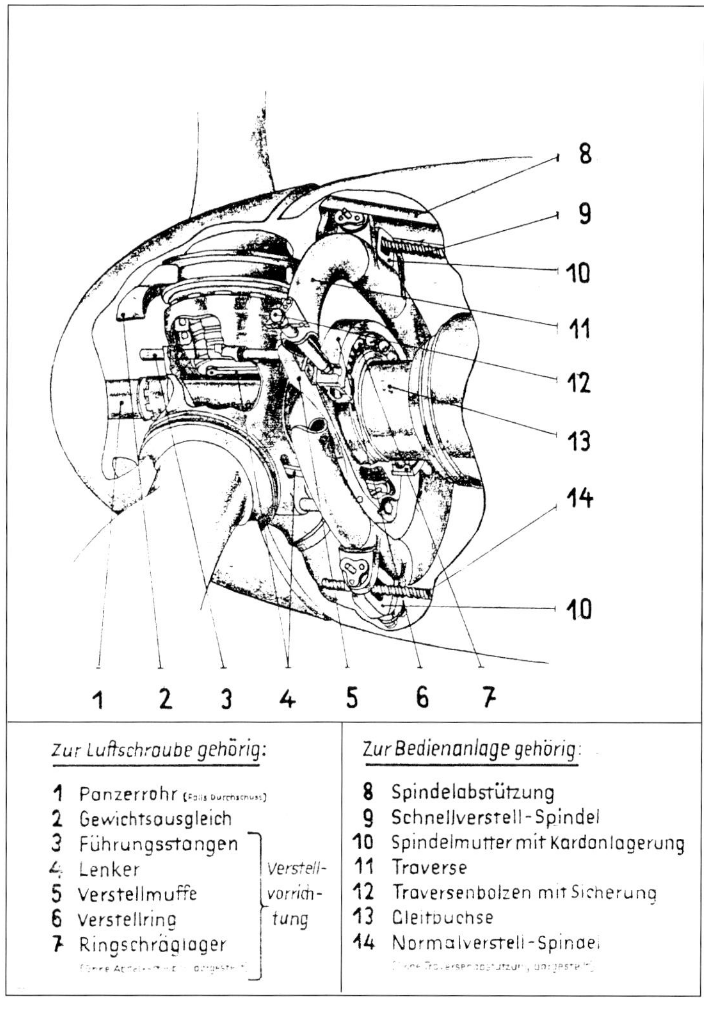


**Drawing depicting the most important components of the MeP 8 variable-pitch propeller.**

### MeP 2

This smallest Messerschmitt variable-pitch propeller was developed in 1940 for engines with outputs of up to 200 H.P. Seven prototypes were built in the Messerschmitt shops, of which the V2 was sent to the *E-Stelle Rechlin* on loan. Previously this propeller had been used by the Research Institute for Motor Vehicles and Aero-Engines in Stuttgart for snow ski experiments. No longer needed at Rechlin, it then went to the Bücker company, which already had the V5 under test. The V2 and V5 were returned to Messerschmitt, where they were repaired and "loaned" to the Gotha company. Siebel also had an interest in the MeP 2 for its Si 202 *Hummel* (Bumblebee).

At Augsburg in June 1941 the MeP 2 was installed on the Bü 181 WerkNr. 0021, NF + OF. In the period from 9 July to 3 October 1941 a total of 145 flights were made with this aircraft totaling 128 hours, 22 minutes.

As it was classified as "non-essential equipment," development of the MeP 2 was not pursued.

## MeP 8

Development of this largest Messerschmitt variable-pitch propeller, which was intended for engines producing 1,400 to 2,000 H.P. and was initially planned for the DB 603 and Jumo 213 power plants, began at the end of 1940. On 7 November 1941 the RLM issued Messerschmitt a contract for 25 MeP 8s. The high-speed pitch motor had an output of 1,200 watts and a pitch change speed of 70 degrees per second.

At a conference held on 28 April 1942 between the RLM, industry advisors, VDM and Messerschmitt, it was decided that VDM and Messerschmitt should together develop a standard variable-pitch propeller based on the Messerschmitt principle to production standard as quickly as possible. In order to carry out this work, it was decided to combine the development departments of VDM and Messerschmitt in Frankfurt/Main. On 11 July 1942 the Messerschmitt team (approximately 30 designers) began moving to Hasselborn. From that point there is no reliable information concerning further development, testing, or any start of series production of the MeP 8.

## MeP 10 and MeP 12

The MeP 10 variable-pitch propeller was a proposal for a hydraulically activated pitch control mechanism, which differed radically from the previous Messerschmitt pitch control principle. The MeP 12 was a counter-rotating version of T*echnische Amt*'s preliminary decision No. 7374, which called for the production of 25 MeP 12s, was reduced to 10 examples on 27 March 1942.

While the RLM admitted that it had no fundamental reservations about the new Messerschmitt variable-pitch system, it did hinder and delay a more extensive test program and saw to it that all development and production was given to VDM, the leading German manufacturer of propellers, at Frankfurt (Main)-Heddernheim.

# Appendix

Nur für den Dienstgebrauch!

# Flugstrecken Bf 109 E/B

Fluggewicht 3,1 t

1. Ausgabe vom 4.7.1940

Einsatzwerte mit taktischem Abzug

Der Generalluftzeugmeister
Technisches Amt
Erprobungsstelle der Luftwaffe Rechlin

**Motor:** DB 601 A
Baureihe 1
Getriebe 1,55
Verdichtung 6,8

**Luftschraube:** VDM
elektrisch verstellbare Steigung
Grundeinstellung . . . 12 h 40 = 25°
Durchmesser . . . . . . . . . . 3100 mm

**Kraftstoff:** 400 l

**Schmierstoff:**

**Abwurfgerät:** 1 ETC 500 kg

**Bewaffnung:** 2 MG. 17
2 MG. FF

**FT-Ausrüstung:** FuG VII a

**Besatzung:** 1 Mann

## Eindringtiefen
BF 109 E
mit 500 kg Zusatzlast

Der Tabelle ist zugrunde gelegt, daß das Flugzeug am Ziel in 3000 m die Kampfhandlung beginnt und in 1000 m beendet. Der Verbrauch für Hinflug enthält die Zuschläge für Anlassen, Rollen, Starten, Steigen auf Reisehöhe, Gleiten (oder Steigen) auf Angriffshöhe. Der Vorrat für den

| Motorangaben | Flughöhe / Belastungszustand | | 1000 m | | |
|---|---|---|---|---|---|
| | | | Vollgas | Dauerleistung | Sparflug |
| Hinflug mit Last | Ladedruck | ata | 1,3 | 1,15 | 0,85 |
| | Drehzahl | U/min | 2400 | 2200 | 1500 |
| | Wahre Geschwindigkeit | km/h | 405 | 370 | 240 |
| | Spez. Brennstoffverbrauch | l/km | 0,81 | 0,75 | 0,73 |
| Rückflug ohne Last | Ladedruck | ata | 1,3 | 1,15 | 0,75 |
| | Drehzahl | U/min | 2400 | 2200 | 1300 |
| | Wahre Geschwindigkeit | km/h | 445 | 410 | 265 |
| | Spez. Brennstoffverbrauch | l/km | 0,74 | 0,67 | 0,45 |

| Entfernung zum Ziel | Hinflug mit Last | | | |
|---|---|---|---|---|
| | Flughöhe | Motorbelastung | Flugzeit | Brennstoffaufwand |
| km | m | | Min. | l |
| 100 | beliebig | Dauerleistung oder weniger | 18 ÷ 23 | 125 |
| 150 | beliebig | Dauerleistung oder weniger | 25 ÷ 32 | 160 |
| 175 | beliebig | Dauerleistung oder weniger | 28 ÷ 33 | 180 |
| 200 | beliebig | Dauerleistung oder weniger | 31 ÷ 53 | 195 |
| 220 | 0 ÷ 5000 | Sparflug | 47 ÷ 58 | 205 |
| 240 | 3000 ÷ 7000 | Sparflug | 45 ÷ 55 | 215 |
| 250 | 5000 ÷ 7000 | Sparflug | 48 ÷ 53 | 215 |

Facsimile of the Bf 109 E service manual with radius of action and information on engine output, fuel consumption and altitudes.

Rückflug enthält die Zuschläge für nochmaliges Steigen auf Reisehöhe, für Restmenge, Gleitflug, Durchstarten. Für Kampfflug am Ziel ist ein Verbrauch von 5 l/min zugrunde gelegt. Strecken bei Windstille. Bei 30 km Gegenwind am Boden (bzw. 50 km in 3000 m Höhe) braucht man 1/5 ÷ 1/6 mehr Brennstoff, bei gleichem Rückenwind 1/10 ÷ 1/12 weniger. Für 5 min Kampf auf dem Rückflug wird die Reichweite um 35 km kürzer.

| 3000 m | | | 5000 m | | 7000 m | |
|---|---|---|---|---|---|---|
| Vollgas | Dauerleistung | Sparflug | Vollg.-Dauer | Sparflug | Vollg.-Dauer | Sparflug |
| 1,3 | 1,15 | 0,85 | 1,15 | 0,85 | 0,92 | 0,69 |
| 2400 | 2200 | 1500 | 2400 | 1500 | 2400 | 1800 |
| 440 | 415 | 270 | 470 | 330 | 450 | 370 |
| 0,8 | 0,7 | 0,7 | 0,69 | 0,63 | 0,6 | 0,48 |
| 1,3 | 1,15 | 0,75 | 1,15 | 0,76 | 0,92 | 0,63 |
| 2400 | 2200 | 1300 | 2400 | 1400 | 2400 | 1600 |
| 490 | 455 | 300 | 520 | 350 | 500 | 360 |
| 0,72 | 0,64 | 0,45 | 0,63 | 0,43 | 0,54 | 0,43 |

| Zeit am Ziel für Kampf und Stürze Vollgasflug | Rückflug ohne Last | | | |
|---|---|---|---|---|
| | Kraftstoffvorrat bei Rückflugbeginn | Flughöhe | Motorbelastung | Flugzeit |
| Min. | l | m | | Min. |
| 35 ÷ 38 | 100 | beliebig | Zulässige Dauerleistung | 16 ÷ 17 |
| | 85 | 0 ÷ 5000 | Sparflug | 20 ÷ 24 |
| 22 ÷ 27 | 133 | beliebig | Zulässige Dauerleistung | 23 ÷ 25 |
| | 105 | 0 ÷ 5000 | Sparflug | 28 ÷ 35 |
| 14 ÷ 20 | 150 | beliebig | Zulässige Dauerleistung | 26 ÷ 28 |
| | 120 | 0 ÷ 5000 | Sparflug | 33 ÷ 41 |
| 8 ÷ 15 | 165 | beliebig | Zulässige Dauerleistung | 29 ÷ 31 |
| | 130 | 0 ÷ 5000 | Sparflug | 37 ÷ 47 |
| 12 | 135 | 0 ÷ 5000 | Sparflug | 40 ÷ 51 |
| 8 | 145 | 0 ÷ 5000 | Sparflug | 44 ÷ 56 |
| 7 | 150 | 0 ÷ 5000 | Sparflug | 45 ÷ 56 |

List of Bf 109 Experimental Aircraft
(only up the E-series)

| Prototype | WerkNr. | Registration | Power Plant | First Flight | Remarks |
|---|---|---|---|---|---|
| A[1] | 757 | | | | Static test airframe |
| V1 | 758 | D-IABI | R.R. Kestrel II S | 28/05/35 | 1936-1941 Stall and Spin tests. Shortened wing slats. |
| V2 | 759 | D-IILU | Jumo 210 A | 12/12/35 | "Fly-off" at Travemünde Feb. 36. Crashed 1/4/36 |
| V3 | 760 | D-IOQY | Jumo 210 B | 08/04/36 | Service trials in Spain with "Legion Condor" |
| V4 | 878 | D-IALY | Jumo 210 B | 23/09/36 | Service trials in Spain with "Legion Condor" |
| V5 | 879 | D-IIGO | Jumo 210 B | 05/11/36 | Trials with EPAD 17 (electric weapons firing) |
| V6 | 880 | | Jumo 210 D | 11/11/36 | Service trials in Spain with "Legion Condor" |
| V7 | 881 | D-IJHA | Jumo 210 G | 05/11/36 | Zurich Air Meet (23/7 to 1/8/38) |
| V8 | 882 | D-IMQE | Jumo 210 D | 29/12/36 | Test-bed for DB 600 at Rechlin |
| V9 | 1056 | D-IPLU | Jumo 210 G | 23/07/37 | Zurich Air Meet (23/7 to 1/8/38) |
| V10 | 884 | D-IXZA | Jumo 210 D | 30/12/36 | "On loan" to MTT-AG. (factory aircraft) |
| V10a | 1010 | D-IAKO | Jumo 210 D | 24/02/37 | 1st test flight with MeP 6 |
| V11 | 1012 | D-IFMO | Jumo 210 D | 01/03/37 | New gun wing with MG 17s |
| V12 | 1016 | D-IVRU | Jumo 210 D | 16/09/37 | New gun wing for MG FF |
| V13 | 1050 | D-IPKY | DB 601 A | 10/07/37 | Zurich Air Meet with DB Racing Motor III, converted into record aircraft |
| V14 | 1029 | D-ISLU | DB 601 A | 28/04/37 | Zurich Air Meet with DB Racing Motor II, crashed by Udet 27/07/37 |
| V15 | 1773 | D-IPHR | DB 601 A | 18/12/37 | Prototype for the E-series. Spoilers. Converted to carrier wing. |
| V15a | 1774 | D-ITPD | DB 601 A | 21/04/38 | Comparison flights with V15 (DB 601 with reduced output) |
| V16 | 1775 | D-IDXG | Jumo 210 D | 10/03/38 | Planned as weapons test-bed. |
| V17 | 1776 | D-IYMS | Jumo 210 D | 24/02/38 | Converted as carrier aircraft. Arrested landing trials at Travemünde. |
| V17a | 301 | D-IKAC | Jumo 210 D | | License built by Erla, from 25/07/38 with Mtt.AG. |
| V18 | 1731 | D-ISDH | Jumo 210 D | 04/02/38 | On 23/02/38 to Rheinmetall-Borsig for installation of weapons |
| V19 | 1720 | D-IVSG | Jumo 210 G | 15/11/37 | Prototype for C-series. |
| V20 | 1779 | D-ICZH | Jumo 210 G | 06/10/38 | To Rechlin on 2/3/39 |
| V21 | 1770 | D-IFKQ | Twin Wasp | | Flew August 1939, Test-bed for Twin Wasp two-row radial engine |
| V22 | 1800 | D-IRRQ | DB 601 A | 14/01/39 | Converted to "F-1", to Rechlin on 21/11/39 |
| V23 | 1801 | D-ISHN | DB 601 A | 09/02/39 | Prototype for the F-series, steerable tailwheel installed. |
| V24 | 1929 | D-ITDH | DB 601 A | 04/03/392[1] | Wind tunnel tests at Chalais-Meudon in mid-1941 |
| V25 | 1930 | D-IVCK | DB 601 A | 04/03/39[2] | Converted into "F-2", To Tarnewitz on 23/11/39 |
| V26 | 1930 | CA + NK | DB 601 A | | E-3/B, to Rechlin on 6/6/40 |

[1] The static test airframe is designated "A" in a Messerschmitt document.
[2] Date of first flight not known, accepted by BAL Augsburg.

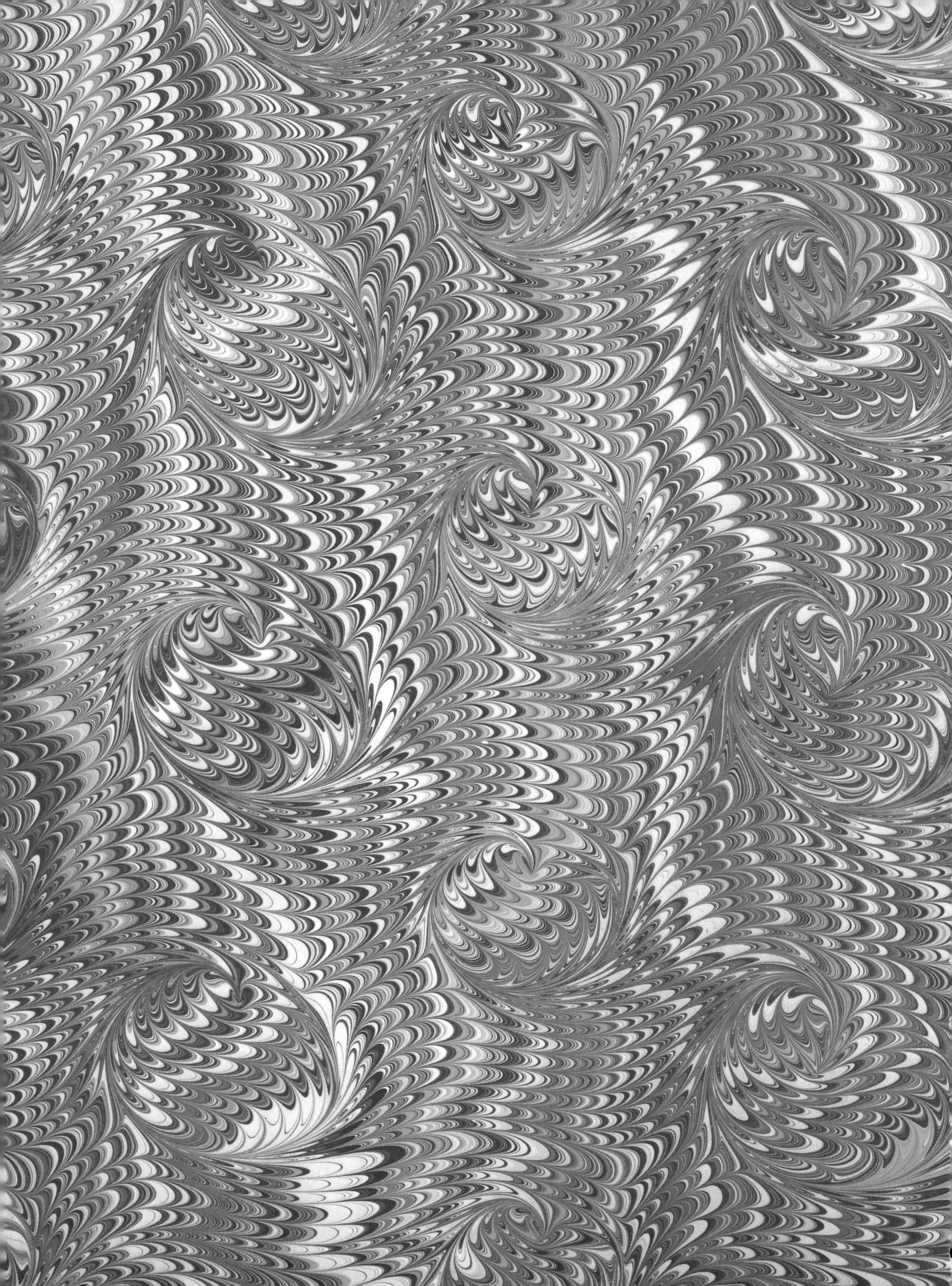